Manual

KOI
HEALTH

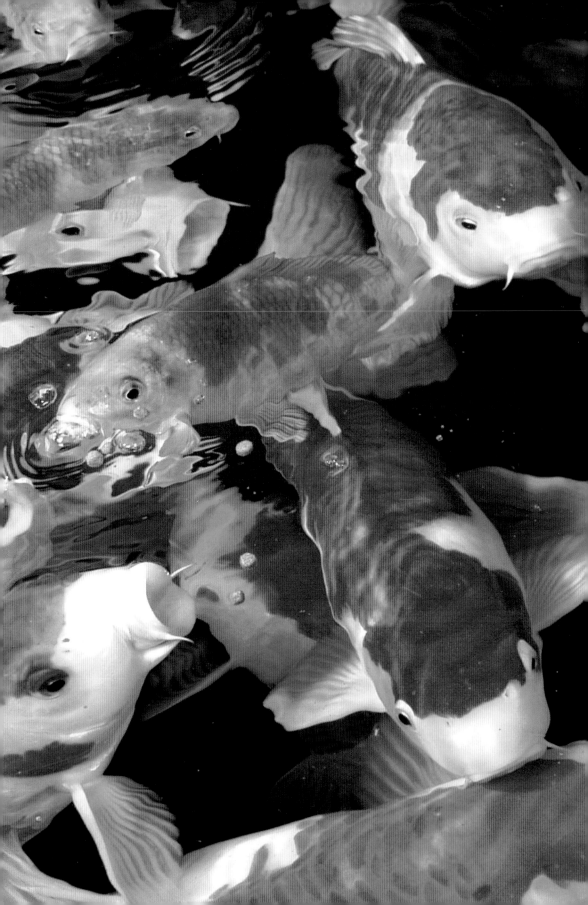

Manual of
KOI
HEALTH

Keith Holmes and Tony Pitham

Consultants: Dr Peter Burgess,
Dr Paula Reynolds, William Wildgoose,
Bernice Brewster

FIREFLY BOOKS

A FIREFLY BOOK

Published by Firefly Books Ltd. 2011

First printing

Publisher Cataloging-in-Publication Data (U.S.)
Holmes, Keith.
  Manual of koi health : how to create a healthy environment for your koi and how to treat any sicknesses that may afflict them / Keith Holmes and Tony Pitham ; illustrated by Neil Sutherland.
2nd ed, rev and updated.
[160] p. : ill. (some col.), photos. (chiefly col.) ; cm.
Includes index.
Summary: Comprehensive descriptions of the common koi diseases and advice on how to recognize and treat such diseases.
ISBN-13: 978-1-55407-920-9 (pbk.)
1. Koi – Health. 2. Koi – Diseases. I. Pitham, Tony. II. Sutherland, Neil. III. Title.
639.3/7483 dc22 SF458.K64H656 2011

Library and Archives Canada Cataloguing in Publication
Holmes, Keith, 1978-
  Manual of koi health : how to create a healthy environment for your koi and how to treat any sickness that may afflict them / Keith Holmes and Tony Pitham.
2nd ed, revised and updated
Includes bibliographical references and index.
ISBN-13: 978-1-55407-920-9
1. Koi--Health. 2. Koi--Diseases. I. Pitham, Tony II. Title.
SF458.K64H65 2011   639.3'7483   C2011-901365-7

Published in the United States by
Firefly Books (U.S.) Inc.
P.O. Box 1338, Ellicott Station
Buffalo, New York 14205

Published in Canada by
Firefly Books Ltd.
66 Leek Crescent
Richmond Hill, Ontario L4B 1H1

Printed in China

Editor: Philip de Ste. Croix
Designer: Philip Clucas MSIAD
Studio photography: Neil Sutherland
Artwork: Phil Holmes and Stuart Watkinson
Production management: Consortium,
Poslingford, Suffolk

The Authors

Keith Holmes is manager of Koi Water Barn. He has worked in the general aquatic industry with both tropical and coldwater species. Holmes is a monthly contributor to numerous specialist aquatic magazines and has published articles around the world. He is also co-author of *A Practical Guide to Building and Maintaining a Koi Pond.*

Tony Pitham is the owner of Koi Water Barn and is one of the most respected koi dealers in Japan. He has judged and photographed the prestigious All Japan Shinkokai show, and has supplied many show-winning koi. He write's for many of the top koi magazines worldwide and is co-author of *A Practical Guide to Building and Maintaining a Koi Pond.*

The Consultants

Dr Peter Burgess is a fish health consultant and university lecturer, specializing in ornamental fish. He holds degrees in parasitology, microbiology and fish biology and was awarded a PhD from Plymouth University for research on whitespot disease in tropical marine fish. Dr Burgess writes regularly for several aquarium and koi magazines.

Dr Paula Reynolds is an aquatic patho-biologist and owner of Lincolnshire Fish Health Laboratories and Research Centre. As well as offering a fish health service to veterinarians, koi dealers and hobbyists, her centre carries out health screening on koi from around the world. Epidemiology, the science of disease transmission, is one of her prime areas of research.

William Wildgoose graduated from Glasgow Veterinary School in 1977 and has worked in London since then. He obtained his Certificate in Fish Health and Production from the Royal College of Veterinary Surgeons in 1997. He has provided professional fish health services to private hobbyists, retailers and importers and has been the veterinary advisor to the Ornamental Aquatic Trade Association for several years.

Bernice Brewster is a private fisheries consultant, with many years experience working with ornamental and coarse fish husbandry. Brewster writes articles for popular fishkeeping magazines and contributes to the study of fish parasitology.

Authors' Acknowledgements

The authors would like to thank – the late John Pitham, Sakai Fish Farm (Hiroshima), Narita Koi Farm (Nagoya), Ogawa fish farm, Nakamori and Co, Michel Capot (Koi Ichi Ban), Peter Burgess, Paula Reynolds (Lincolnshire Fish Health Consultancy), *Rinko* magazine, Martin Plows, the Koi Water Barn team, the Ornamental Aquatic Trade Association (OATA), Lisa Holmes for her patience during the writing of this book, and everyone else who has helped to make its publication possible.

CONTENTS

Note To The Reader

While every effort has been made to ensure that the content of this book is accurate and up-to-date at the time of going to press, it must be accepted that knowledge of fish diseases, the side-effects of treatments (on both fish and humans), and other aspects covered, is an ever-changing science. Moreover, the manner in which the information contained herein is utilised is beyond the control of the authors and publisher, although every effort has been made to warn of known dangers. For these reasons no responsibility can be accepted by the authors and publisher for any loss, injury, or other problem whatsoever experienced by any person using this book.
Always seek the advice of a veterinarian or a qualified koi health specialist before applying the treatments described in this book.

FOREWORD

by Dr Peter Burgess
Researcher and lecturer in ornamental fish health and husbandry

As with any pet, koi are susceptible to a variety of ailments that may crop up from time to time. Fortunately, many common health problems of koi can be successfully treated, provided the koi keeper has the basic knowledge to be able to diagnose and deal with the problem at an early stage. The aim of this book is to arm the koi keeper with this essential knowledge.

The Interpet Manual of Koi Health has been written specifically for koi keepers, whether they are novices with just a single pond in the back garden, or serious koi enthusiasts who are deeply involved in breeding or exhibiting their fish. In contrast to many textbooks on fish health, this manual presents technical information at a level that is readily understood by the non-scientist. It is primarily designed to be a practical guide, based largely on the authors' first-hand experiences gained over many years of working in the koi industry. As such, the information herein reflects current practices within the professional koi community and will be of interest to those employed in this field. The book is arranged in a logical sequence comprising four parts.

Part 1 addresses the all-important subject of disease prevention. The key to healthy koi-keeping lies in providing optimal environmental conditions and a well-balanced diet. With this aim, the reader is introduced to essential topics such as water chemistry and water quality, filtration systems, and koi nutrition. Maintaining a healthy koi pond is similar to looking after a car: various checks and services have to be performed at recommended intervals. These routine tasks are clearly laid out on pages 20-21. In terms of koi health, the old maxim "prevention is better than cure" is extremely apt, and for this reason it is recommended that Part 1 be read in its entirety.

Part 2 covers the "essential skills" that the koi keeper will need to acquire in order to monitor the health of his stock. This section begins with a simple overview of koi anatomy and biology, followed by practical tips on how to net and handle koi correctly, how to transport them safely, and how to use a microscope to check for parasites. Specialist procedures are also described, such as the taking of tissue samples for disease identification and laboratory investigations.

Part 3 comprises the major section of the book, dealing with commonly encountered koi health problems. When faced with a disease outbreak, the reader is recommended initially to study the various parasites and disease conditions that are illustrated on pages 58-61. These four pages serve as diagnostic guides to the various viral, bacterial, parasitic and other disease conditions that follow in A-Z order. Beginning with *Aeromonas* infections, each A-Z entry contains vital information regarding the disease, how to recognize it, and how to prevent and treat the condition. Where

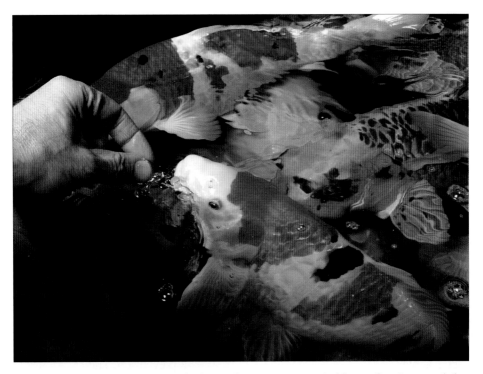

chemical treatments are required, the authors suggest suitable medications and dose rates, plus other treatment information that will improve the chances of a successful cure. Each entry is lavishly illustrated with colour photographs depicting characteristic symptoms of the disease in question, thereby helping the koi keeper to arrive at a correct diagnosis. Where appropriate, there are close-up pictures of various parasites and easy-to-follow diagrams showing the often complex and fascinating life cycles that some of these parasites employ.

Part 4 describes various methods for administering treatments to koi, such as medicated baths and dips and the surface treatment of wounds. Here the reader will find additional information on chemical treatments in current usage. Some "advanced" techniques are also described, including methods for sedating koi, and the delivery of antibiotics by injection. It should be emphasized that these procedures (including skin scraping techniques described in Part 2) should not be performed without proper training. Ideally, they are best left to a veterinarian or suitably qualified koi health specialist.

Finally, it is important to mention that koi (and other fish) are capable of experiencing stress and there is mounting scientific evidence that they can perceive pain too. With this in mind, it is the responsibility of every koi keeper to ensure that his or her fish are given the same high level of care as is afforded to any other pet. The valuable information contained within this book will assist greatly in achieving this goal.

PREVENTING DISEASE

The best way of preventing disease is to control and limit the levels of stress to which your koi are exposed. Stress can be dramatically reduced if you have a basic understanding of water quality and realize how each different parameter will affect your fish. This allows you to recognize the signs of poor water quality quickly and to react to them in the appropriate way. This should be backed up by regular water testing and the various ways of doing this are described in this section. Steps to remedy poor water quality are also discussed here, and if this advice is followed, many disease problems can be stopped in their tracks by simply maintaining optimum water conditions. The old adage holds true: maintaining the water in a koi pond is like running a small-scale sewage works. If water is kept in optimum condition, other problems will be far less likely to occur.

Maintaining optimum water quality should go hand in hand with a good, regular programme of system maintenance and husbandry. A basic guide to the essential steps which you should observe is also provided in this section. Obviously every pond is different and your pond may require additional maintenance, but the points highlighted here should all be part of your overall maintenance programme. If your schedule and time allow, try to develop a routine whereby a set period of time can be dedicated to your pond every week just to attend to these basic jobs. If they are left undone, you could face potentially serious problems. The overall health of your pond can be further enhanced by the use of heating, specialist filtration equipment, such as ozone, and UV sterilization, and these are all examined here. These items are by no means essential to running and maintaining a healthy pond, but are all worthwhile additions if your system and budget will run to them. Heating, perhaps, is the exception as many koi keepers and koi professionals regard heating not as a luxury item, but as an essential part of a healthy koi pond.

Finally this section considers feeding, as this can have a significant effect on the health of your koi. It is vital that the correct foods are fed at the right time of the year – the reasons for this are outlined, and the use of specialist foods described. Overfeeding can be one of the main causes of poor water quality and guidance will be found as to the correct feeding ratio for your koi. This advice should be integrated into your overall pond health and maintenance regime.

The practices described in this section alone are actually the first line of defence in preventing many

Below: Keeping healthy koi starts with creating and maintaining a stable and healthy environment.

of the more serious health problems which are examined in section 3. If any adverse behavioural changes are noticed, water quality should always be the first port of call. If water quality proves to be acceptable, then the steps outlined in section 2 should be followed to allow you to test for the presence of many diseases. Section 3 will confirm the exact diagnosis and provides treatment advice.

Finally section 4 explains how to carry out specific treatment methods which are referred to throughout this book. So in one compact, easy-to-follow volume you now have a complete health care manual incorporating advice on disease prevention, identification and diagnosis, and the most up-to-date treatment.

THE EFFECTS OF STRESS ON YOUR KOI

As you read this book it will quickly become clear that one of the most important things you can do as a responsible fishkeeper is to reduce the amount of stress to which your koi are exposed. This single precaution could be an absolutely critical factor in reducing the outbreak of health problems among your stock. The cause is simple – when a koi becomes stressed, its defences against disease are weakened and this makes infection more probable.

How can stress most effectively be reduced? The answer is to create a stable environment in which to keep your koi and, wherever possible, to avoid environmental change. However, this is easier said than done as there are numerous factors acting on a fish which all have the same end result of causing stress. Nevertheless, a few basic rules should be followed to help establish

Above: Transportation can be a highly stressful experience for koi and leave them vulnerable to infections.

and maintain the most stress-free environment that you can. There are two key factors to observe – avoid overstocking and maintain optimum water quality. Poor water quality is one of the main causes of stress, and so regular, almost religious, water tests should be carried out; if identified sufficiently early, the necessary steps can be taken to rectify any problem with the water, and so limit the amount of stress to which your koi are exposed.

Basic fishkeeping tasks, such as netting a koi for inspection, are also highly stressful to koi. Netting causes the 'fight or flight' hormone adrenalin to be released into the circulating blood. Adrenalin induces profound physiological changes, notably increased heart rate and blood pressure, dilation of major arteries to body muscle and constriction of blood vessels supplying the skin and digestive system. When netted, it is quite common for koi to eject any food in the upper intestine. This allows it to divert vital energy resources away from digestion and to the muscles that enable it to escape. Another common effect of adrenalin is to cause the koi to bleed profusely from the gills due to the increased blood pressure, causing one or more of the delicate blood vessels to rupture. Such blood loss can be alarming, but fish have evolved a special mechanism whereby water causes the blood to clot. The effects of stress are cumulative; shipping koi from Japan leads to highly stressed koi on arrival at their destination and it takes months for them to recover.

Environmental Changes

Even the process of purchasing a new koi for your pond is highly stressful for the fish as the whole business of netting, bagging and boxing, and the journey to its new home subjects the fish to high stress levels. When this is combined with the potential changes in water temperature, pH and hardness that the koi will experience when transferred to its new pond, a highly stressful situation results. When your koi have to cope with a situation like this during which dramatic environmental changes have occurred, they will naturally try to adjust to the new environmental conditions in which they find themselves. However, this consumes a lot of the koi's reserves of energy as it struggles to adjust to these changes; as a result, other physical functions, including the immune system, may suffer during this period of acclimatization. Your koi have their guard down and are vulnerable to infection. These examples show how difficult it is to limit the effects of stress on your koi. Stress may (and probably will) occur while you are performing even the simplest of routine tasks.

So, what is the answer? To help combat the effects of stress on your koi and reduce the likelihood of opportunistic infections occurring, it is vital that a well-maintained system is kept, and that excellent husbandry skills are employed. Overstocking and unnecessary handling should also be avoided. Keeping a well-maintained pond and observing good husbandry should also speed up the recovery process when diseases do occur, particularly as the illnesses themselves actually add to the stress burden on the koi. Even the process of treating your koi with medications may prove stressful to the fish, and thus optimum environmental conditions are required throughout to aid quick recovery. Stress prevention should form an essential part of your healthcare regime. If stress levels can be kept to a minimum, the likelihood of diseases taking hold will also be diminished.

Below: As a koi keeper, try to avoid excessive use of the net as it will have an effect on how stressed your fish feel.

Stocking a Koi Pond

For many years the recommendation for stocking a pond was based on length of fish in relation to the surface area of the pond, i.e. 7.5cm of fish length, including the tail, for every square metre of water surface. This formulation was far from ideal as it gave no consideration to the water depth. The preferred method of stocking a koi pond is to allow a maximum of 8kg fish weight for every cubic metre of water volume.

Large numbers of koi in a confined pond is stressful. Not only are the koi competing for individual space but also for limiting resources, most notably dissolved oxygen concentration. Infectious disease, such as Koi Herpesvirus (KHV), has a much greater impact on heavily stocked ponds as the koi tend to be stressed through overcrowding. As a result the immune system is compromised, allowing diseases to spread rapidly through the population. In the event of a parasite infection, the balance is tipped strongly in favour of the parasite, as it can readily find new hosts and will spread quickly to infect the koi.

A lightly stocked pond is much easier to manage than a heavily stocked one, but realistically most koi keepers are reluctant to admit their pond holds too many koi.

WATER QUALITY

To help to prevent outbreaks of disease in your pond, it is vital that water quality is maintained at the highest possible level. Most illnesses become more serious when koi are stressed, and poor water quality is one of the biggest stress factors that can occur within the pond. It is easy to test for the main water quality parameters – pH, ammonia, nitrite, nitrate, and carbonate hardness (KH) – and these can be easily rectified should a problem arise. Koi produce nitrogenous waste in the form of ammonia, which is the main source, although decaying organic material may also contribute a small amount. *Nitrosomonas* species of bacteria in the biological filter will oxidize the ammonia into nitrite. The next stage involves nitrite being converted into nitrate by *Nitrobacter* bacteria. Both ammonia and nitrate are taken up by plants as a source of natural fertilizer and this is why many koi ponds incorporate some form of vegetable filter to remove these nitrogen products from the water and reduce the growth of nuisance

Above: Testing your water should be a key part of your koi healthcare regime. Test kits, like this strip which changes colour when dipped in the water, make this simple to do.

algae. These processes all require the presence of oxygen and are referred to as aerobic. However, if little or no oxygen is present, the *Nitrobacter* and other microbe species will convert nitrate back to nitrite, which may lead to pollution of the koi pond. Denitrification filters exploit the process to reduce nitrate to nitrogen gas. In a pond with a good filtration system, good water quality will

How the nitrogen cycle works

Water returning to the pond from the last stage of the filtration system may still contain some nitrate. Nitrate is one of the nutrients responsible for promoting the growth of blanketweed.

Water changes are an important aspect of pond management, especially if ammonia or nitrite are polluting the pond. Regular water changes can also help to reduce nitrate concentration in the water.

By adding an oxygen atom into each molecule, aerobic bacteria (*Nitrobacter* spp.) convert nitrite into nitrate (NO_3). Nitrate is the final breakdown product of ammonia in the nitrogen cycle. It is far less toxic than ammonia or nitrite.

Protein present in food is used by koi for tissue repair and maintenance, growth and reproduction. Any excess protein cannot be stored and is excreted as ammonia. The protein in any uneaten food also ends up as ammonia.

By removing the hydrogen and adding oxygen into each molecule, aerobic bacteria (*Nitrosomonas* spp.) convert ammonia into nitrite (NO_2). Although not as harmful as ammonia, nitrite is still poisonous to koi.

Ammonia (NH_3) is released into the water via the fishes' gills. The small amount of urea voided in dilute urine also breaks down to form ammonia. Ammonia is very poisonous to koi.

be maintained by the biological filtration system. However, it is important to observe a routine in which at least the levels of pH, ammonia, nitrite and nitrate are regularly tested as small changes in water temperature, an increase in stocking, or even a change in food can all have a detrimental effect on the efficiency of the biological filtration, which ultimately will cause poor water quality.

There are many test kits available that make it easy to test pond water. They come in many forms: liquid, tablet and even strips of testing paper that are simply dipped into the pond water and left for a period of time before the change in colour is compared against a chart. These tests are generally inexpensive, and should be used at least once a month, although weekly or fortnightly is better. If you require more accurate readings, electronic testers are available, but generally at a much higher cost. In fact, although these give a more accurate result, the reading from tablet or liquid tests is generally a good enough indication for the average koi keeper. It is vital that any adverse readings should be acted upon immediately as failure to do so will only prolong the stress to your koi. Furthermore, long exposure to poor water quality can cause quite serious health problems on its own, as each parameter will affect your koi in a different way.

pH

This is the measure of acidity and alkalinity. pH 7 is neutral, below 7 is acid and above 7 is alkaline. The pH measurement is logarithmic, which means a change of a single unit, say from pH 7 to pH 8, is actually a tenfold increase. This is why even a small change in pH can have quite dramatic effects on your koi, although mostly they will tolerate brief changes in pH, such as may be caused by algal photosynthesis. Koi prefer a pH range of 6.5–8.5. Fluctuations in pH are largely buffered by the carbonate hardness (KH) of the water. Extreme changes in pH of the water affect the biochemistry of the blood; and if the pH value of the pond water becomes either too acidic or alkaline, falling outside the acceptable range, koi will start to show symptoms of acidosis or alkalosis. Acidosis occurs when the pH level drops below 5.5 and very acidic conditions result. Following a rapid drop in pH, koi will start to behave erratically and try to escape from the pond by jumping. You will also observe increased gill movements, which result

Broad range pH testing

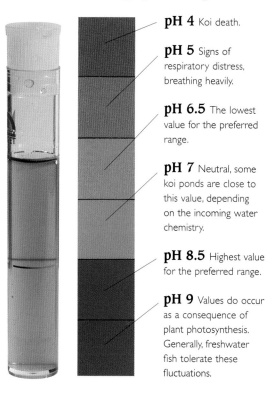

pH 4 Koi death.

pH 5 Signs of respiratory distress, breathing heavily.

pH 6.5 The lowest value for the preferred range.

pH 7 Neutral, some koi ponds are close to this value, depending on the incoming water chemistry.

pH 8.5 Highest value for the preferred range.

pH 9 Values do occur as a consequence of plant photosynthesis. Generally, freshwater fish tolerate these fluctuations.

Above: It is vital to test the pH of water routinely as this affects the toxicity of ammonia.

in fish becoming very stressed, leaving them susceptible to infection. If left untreated, losses may occur just from the change in pH level. If the pH change is much more gradual, with the water becoming very acidic over a long period of time, you may not notice immediate changes in your koi's behaviour. Over time, however, the ability of the gills to extract oxygen from the water will decrease because colloidal iron is being deposited, and this will be indicated by gasping and irregular gill movements, as well as the production of excess mucus and the appearance of areas of redness on the skin. To reverse the effect of acidosis the pH must be brought back to an acceptable level. Avoid the temptation of doing this rapidly as a sudden increase back to correct levels can be just as stressful as doing nothing. To make the water more alkaline, crushed oyster shells placed in the filters are a good remedy. Alternatively you may wish to consider using a proprietary product designed to increase pH value.

Alkalosis occurs when the pH value rises above 9 and the water becomes too alkaline. Many of the symptoms associated with alkalosis are the same as those for acidosis, but koi may also show signs of severe fin and gill damage if exposed to high pH levels for any prolonged period. To lower pH, use aquatic peat in the filters or an off-the-shelf pH-lowering product. Never change pH value too quickly, as this can be very stressful to koi. Once the correct pH value is achieved, you must establish why it changed in the first place. Typical causes of high pH include the presence of lime in the water, which may have leached out from rocks or cement around the pond. The commonest cause of high pH is photosynthesis, either due to submerged plants or nuisance algae such as blanketweed or green water. During daylight, plants produce sugars, using energy from sunlight and carbon dioxide, a process known as photosynthesis. As plants and algae photosynthesise, they consume dissolved carbon dioxide. Water with dissolved carbon dioxide is acidic, so as the plants remove carbon dioxide from the water, the pH rises and the water becomes more alkaline. At night, when photosynthesis stops and respiration dominates, carbon dioxide is released into the water, causing it to become more acid and the pH to drop.

General Hardness (GH)

This is the total amount of salts and minerals present in a body of water. This quantity is basically made of temporary hardness, which is removed by precipitation when water is boiled (it is otherwise known as carbonate hardness), with the remainder consisting of permanent hardness from the salts and minerals that remain in solution after boiling. As most salts and minerals fall into the category of carbonate hardness (KH), koi keepers test for this category more frequently. Different countries use different scales to measure general hardness, so it is vital to check which scale the test kit is using.

Carbonate Hardness (KH)

This is the measure of calcium and magnesium carbonate and bicarbonate present in water and contributes to the water alkalinity. Carbonate hardness plays a vital role in buffering the changes in pH associated with photosynthesis. Soft water tends to be associated with acidic conditions, while hard water is indicative of

Carbonate hardness (KH) test

1: Add the KH reagent a drop at a time to a 5ml sample of pond water, and gently swirl the tube to mix thoroughly.

2: Initially the sample turns blue.

3: As more reagent is added, the water sample turns yellow. Continue counting the drops added until the yellow colour is stable. Each drop added during the test represents 1°dH, which is equivalent to 17.5mg per litre of carbonate.

alkaline water. If you experience pH fluctuations, it is worth testing the pond water for hardness. If you are experiencing low levels of carbonate hardness, you may need to add minerals, as koi need a certain amount of calcium and minerals for good health. They help to buffer the water, thus maintaining a constant pH level. Use crushed oyster shells in the filters, or add a proprietary buffering product for ponds. Certain medications react differently in soft or hard water, so it is important to know about your water conditions should a problem arise. Generally in soft water chemicals become more toxic, while in hard water they become less so.

Ammonia (NH_3/NH_4)

This is released into the pond in the form of waste from your koi and from decomposing organic matter. Ammonia exists in two forms in water, free ammonia (NH_3) and ionized ammonia (NH_4). Free ammonia is the more harmful and the effect it has on koi increases with the temperature and pH of the water. Higher pH values make the ammonia more toxic. Common salt used in the pond will have the opposite effect and decrease the toxicity of ammonia. Most test kits just give one reading for total ammonia

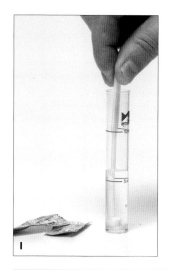

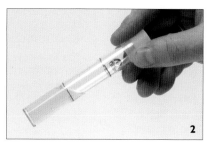

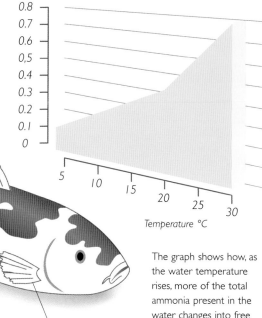

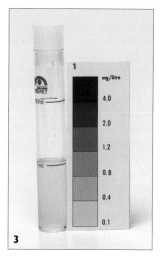

1: Here a tablet test for ammonia is being used and the tablets are being crushed.
2: Once the tablets are dissolved, the tube must be shaken to mix the contents.
3: After a set period of time, the water will change colour and this can be compared against the colour chart supplied with the kit.

which includes free and ionized ammonia, but it does indicate whether an ammonia problem is present or not. Koi kept in water with high levels of ammonia experience numerous problems,

Ammonia toxicity and temperature

% free (toxic) ammonia in total ammonia

[graph with y-axis values 0, 0.1, 0.2, 0.3, 0.4, 0.5, 0.6, 0.7, 0.8 and x-axis Temperature °C with values 5, 10, 15, 20, 25, 30]

Temperature °C

High ammonia levels make koi susceptible to other conditions, such as fin rot.

The graph shows how, as the water temperature rises, more of the total ammonia present in the water changes into free ammonia which is very toxic to the koi in the pond.

including the build-up of cells on the surface of the gill lamellae. This reduces water flow over the gills and thus the amount of oxygen that can be absorbed. The condition is known as hyperplasia.

Removal of ammonia from the koi body is passive and relies on a lower concentration of ammonia in the water to release it at the gills. If the water contains ammonia, this substance accumulates in the body and tissues of the koi and alters the pH of the blood, which makes it difficult for the red pigment, haemoglobin, to pick up oxygen. Ammonia is also a potent neurotoxin, damaging the nervous system.

To reduce high ammonia levels, carry out regular water changes of 10-20 per cent on a daily basis in extreme cases. Filter-boosting products may be required to improve the performance of the biological filtration. Use purified tapwater for the water changes or remove chlorine and other harmful substances from the water with a dechlorinator. Zeolite can also be added to the filters. This rock naturally absorbs ammonia over time before it needs to be recharged with salt. Once the correct levels of ammonia are established, investigate why they rose so high. Common causes include introducing new stock, the filter becoming blocked with solid waste, pump failure, overstocking, overfeeding, and even over-zealous filter maintenance. Cleaning filters with tapwater, for example, may kill some or all of the beneficial bacteria, thus impeding biological filtration. The other common cause is a new set-up where the filters have yet to mature.

15

Nitrite (NO$_2$)

Nitrite occurs naturally in a koi pond as ammonia is broken down by bacterial action. Although not as toxic as ammonia, nitrite is poisonous to your koi if levels are allowed to become too high. In water with a high nitrite level, the koi may start to jump, or flick against the sides of the pond as if being irritated by something. This may appear to indicate a parasite or other infection, as these symptoms are common in many such diseases. However, first test for nitrites. Nitrite binds with the oxygen-carrying haemoglobin molecules in the koi's blood cells, forming methaemoglobin which does not carry oxygen. As a result the blood and gills take on a brown appearance. If left, high nitrite may result in losses of koi, although some koi do tolerate high nitrite levels better than others. Nitrite is a skin irritant, causing koi to rub and flick, but it also affects the smooth muscles, leading to cardiovascular collapse and death. The addition of salt to a pond can help reduce the toxic effects of nitrite, but the best course of action is to follow the same procedure for reducing high ammonia. High nitrite levels may be associated with immature filters or low dissolved oxygen in the pond and filters.

Water Quality Criteria: Recommended Levels

The following table shows the recommended maximum/minimum levels for various chemicals present in water. Outside these levels, remedial action should be taken.

Dissolved oxygen	min 7-8mg/l
Free ammonia	max 0.02mg/l
Nitrite (in soft water)	max 0.01mg/l
Nitrite (in hard water)	max 0.02mg/l
Nitrate	max 50mg/l above ambient tapwater

A vegetable filter

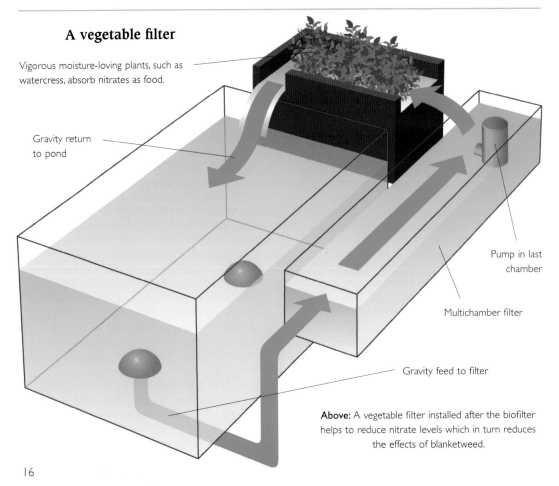

Vigorous moisture-loving plants, such as watercress, absorb nitrates as food.

Gravity return to pond

Pump in last chamber

Multichamber filter

Gravity feed to filter

Above: A vegetable filter installed after the biofilter helps to reduce nitrate levels which in turn reduces the effects of blanketweed.

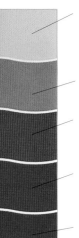

Less than 0.3mg/litre Exceeds the recommended maximum exposure for both soft and hard water. Make partial water changes and monitor regularly.

0.3mg/litre In soft water nitrite is more toxic to koi. Make partial water changes.

0.8mg/litre At this level nitrite is harmful to koi in both soft and hard water. Make regular partial water changes.

1.6mg/litre Nitrite combines with red blood pigment affecting ability of koi to use oxygen. Make daily water changes.

Over 3mg/litre Nitrite pollution now a serious problem and can kill fish. Water changes are essential.

Above: Nitrites are toxic to koi, so it is essential to use test kits to check the nitrite level in your water regularly.

Nitrate (NO_3)

This is produced as the last stage of the nitrification process, when nitrite is converted to nitrate. Although nitrate is apparently tolerated by koi – and is certainly less poisonous than either ammonia or nitrite – there is now experimental evidence that nitrate causes stress to koi. The effect of stress through exposure to nitrate results in susceptibility to infectious disease and affects growth and reproduction, fry and young koi being especially sensitive. High nitrate concentrations also reduce the available dissolved oxygen, which increases the stress placed on the koi. Nitrate can be reduced through regular partial water changes but many koi keepers opt to have a vegetable filter with attractive marginal aquatic plants. Trickle, wet/dry or shower filters will reduce the concentration of nitrate. Alternatively, a denitrification unit may also be used to control nitrate levels.

Oxygen (O_2)

Oxygen enters the pond by the process of diffusion at the surface and at the same time carbon dioxide is released from the pond. If plants are present, the process of photosynthesis causes oxygen to be released into the water, although this is reversed at night when oxygen is absorbed by plants and carbon dioxide is given off. Oxygen levels are dramatically affected by

temperature – the higher the temperature, the lower is the dissolved oxygen content of the water. This creates a Catch 22 situation as koi most need oxygen in the hot summer months just when the water is least able to hold high levels of oxygen. As a guide, a minimum level of 7-8mg/l of dissolved oxygen should be maintained if the temperature will allow, although at temperatures of 30°C (86°F) and above it may prove impossible to maintain these levels. It is

How temperature affects oxygen levels

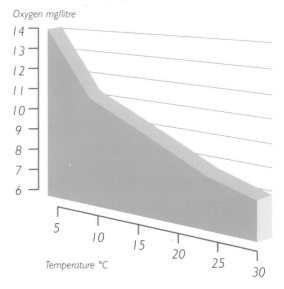

Oxygen mg/litre

Temperature °C

Above: Oxygen solubility decreases with rising temperature. Affected koi are initially lethargic but as oxygen levels fall, they rise to the surface gasping.

Gill cover flares as the koi tries to extract as much oxygen as possible from the water.

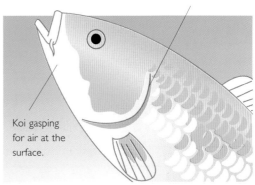

Koi gasping for air at the surface.

Above: Dissolved oxygen should be monitored, especially in hot weather: here an electronic DO tester is being used.

important to make oxygen tests on site at the pond, otherwise a false reading may be produced. If koi become lethargic and stop feeding in the summer, this is usually the first sign of low

dissolved oxygen. By the time they are at the surface and gasping, they are within minutes of dying. If low dissolved oxygen is suspected, increase the aeration in the pond by adding water pumps to cascade water, air pumps, fountains or anything that causes considerable disturbance to the water to increase the surface area available for gas exchange. It is worth investing in either an electronic oxygen meter or suitable test kits to monitor this vital gas in the summer. It is worth remembering that it is not just your koi that require oxygen, the bacteria in your biological filter do too, as do the many organisms found in the pond. When low dissolved oxygen levels are experienced, koi will stop feeding as they are unable to digest. The koi will only start feeding again when the dissolved oxygen has reached an optimum. Oxygen depletion may occur for numerous reasons, such as overstocking, insufficient aeration, poor system maintenance, excessive temperatures, and algae blooms. In

Oxygen in the pond – winter

Oxygen is readily soluble in cold water.

The fish are producing less waste and the lower temperatures inhibit bacterial activity.

The koi are inactive at low temperatures and their oxygen requirement is minimal.

Oxygen in the pond – summer

Blanketweed produces oxygen during the day, but consumes it at night.

Less oxygen is available in the water as the temperature increases.

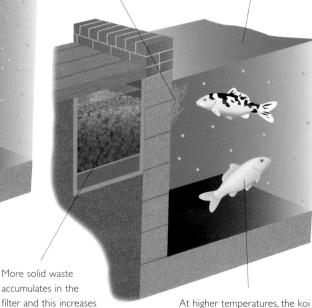

More solid waste accumulates in the filter and this increases the demand for oxygen by the aerobic bacteria.

At higher temperatures, the koi are increasingly active and need more oxygen.

exceptionally cold spells oxygen problems may also result if your pond is allowed to freeze over completely as it will seal the pond's surface from exposure to the air and prevent diffusion from occurring. Ideally your pond should be heated to prevent freezing (see pages 26-7 for more information on pond heating), or if this is not possible it should be at least covered to protect it from the elements, and a small floating heater installed which will keep an area free of ice.

Temperature

This has a dramatic effect on koi as the lower the temperature the slower their metabolism will function. At very low temperatures their immune system will also be affected. Koi are coldwater fish, but they do benefit from being kept in a heated pond with stable water temperatures. In unheated ponds temperature fluctuations cause stress, and koi are highly susceptible to disease over the autumn-winter-spring seasonal change. As the water temperature falls below 14°C (58°F), a koi's immune system becomes less efficient, while at temperatures below 10°C (50°F) it is to a large extent shut down. This is a dangerous situation, as many pathogens are still active. Medications are least toxic and best used at low temperatures but are most effective when the koi's immune system is fully functional. In addition, as the temperature drops so the koi's metabolism slows and its need for food decreases, resulting in the fish being far weaker than they are in the hot summer months.

If the water is allowed to cool, your koi will become torpid, remaining largely inactive, occasionally swimming leisurely around the pond but not feeding. The swimbladder, a buoyancy organ, keeps the resting koi afloat, so they are not lying on the hard, pond base. Young koi under two years old may appear as if they are dead in cold water, and sometimes actually lie on the bottom of the pond or simply float in the water, and will only move if provoked with a net. This makes them highly susceptible to numerous infections. The condition is known as sleeping sickness, and is described on page 129.

These are the main water quality parameters which should be tested on a regular basis; however, it is worth mentioning that the presence of metals, chlorine, chloramines, pesticides, phosphates and other chemicals can also have an effect on your koi. These generally enter the

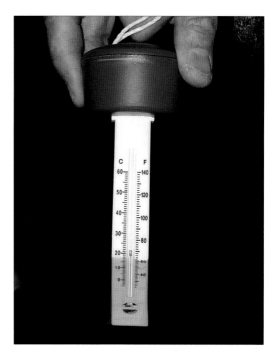

Above: There are several types of thermometer available that allow the koi keeper to monitor the water temperature routinely. This is a simple floating model.

pond when new tapwater is added, and they do not need to be present in particularly high levels to cause a problem. It is vital that all tapwater should be conditioned before it enters the pond, either by the use of a chemical dechlorinator, or by passing it through a water purifier. These are available in numerous sizes, and can be purchased to remove more metals.

To determine if a specialist unit is required for metal removal, it is worth obtaining a water analysis, which can normally be provided by your water supplier. Metals may also enter the pond if any unsuitable metal pipework has been used, so it is vital to ensure that all equipment installed around the pond is suitable for use in an aquatic environment that will contain fish. Many garden ornaments and copper-based fountains are made of, or contain, harmful metals, so take great care if you install such items in your pond.

One final warning – the use of chemicals in the garden, particularly insecticides and pesticides which can be carried on the wind, carries risk as they can be toxic to fish. Ideally garden organically if you keep koi.

GOOD HUSBANDRY & SYSTEM MAINTENANCE

In order to maintain optimum water quality it is vital that a proper maintenance regime is followed. One of the most significant factors contributing to outbreaks of disease is the build-up of waste and sediment within the pond. The removal of this is central to any good system maintenance programme. To ensure that all jobs are carried out when required, it is a good idea to develop a timetable which can be followed throughout the year. It should include the following tasks:

Daily tasks

- Check all koi for signs of behavioural change or physical damage. The ideal time to do this is when feeding, as when the fish are feeding from the surface you will not only be closer to the koi but you will also be able to see underneath the fish.
- Feed all koi, the frequency of this depends upon the time of year (see pages 22-5).
- Check pumps, air pumps, UV filters etc. to ensure that everything is functioning.
- In the summer months you should discharge your filter systems on a daily basis as your koi will be eating more heavily, and so larger volumes of waste will accumulate in the filters
- If your pond is a pump-fed system, it may be necessary to clean any pre-filters on the pump on a daily basis to ensure sufficient flow gets through to the filter.
- In the summer blanketweed may be a serious problem. Its growth can be so rapid that daily removal is required to prevent the pond from becoming clogged with weed, and so restricting water flow from either the bottom drain or pump.

Weekly tasks

- In a gravity-fed system, it is advisable to purge all bottom drain pipework at least once a week in the summer. This is done by shutting the valve to the first chamber of your filter (normally a 10cm/4in slide valve or ball valve), and allowing this chamber to empty by opening its discharge valve. When empty, the slide or ball valve can be re-opened; this will cause a massive surge of water to pass through the drain pipework and flush out any debris which has settled out in the pipework.
- If a skimmer is installed, it is important that it is checked at least once a week, and the collection basket emptied. At certain times of the year, particularly autumn, it may be necessary to do this daily to prevent the skimmer from clogging which will stop leaves from being removed and so allow them to decompose in the pond.
- On both pump-fed and gravity-fed systems, any mechanical filter media, such as brushes or foam, should be cleaned at least once a week to allow a good flow of water through the filter.
- Although not essential, it is certainly a good idea to set aside 30 minutes each week to test the water quality in the pond, and to take any remedial steps that may be necessary (see pages 12-17).
- Any koi which display physical damage or infections which require topical treatment should be netted and treated at least once a week. In some cases you may need to treat twice a week until enough improvement is seen for the frequency of treatment to be decreased.

Above: Skimmers remove surface debris, like food and leaves, from the pond; if not cleaned, they start to block.

Above: Be sure to clean your skimmer on a regular basis by simply washing it in a bucket of water or with a hose.

Monthly tasks

- If your pond is heated, it may be necessary at certain times of the year for you to adjust the heating system to raise or lower the water temperature. When doing this, only small changes of one or two degrees every few days should be made.
- As the seasons pass from autumn into winter, any aeration in the pond should be decreased to limit the effect of water chilling by reducing the amount of cold air which can diffuse into the pond. Also turn down (or off) any waterfalls and fountains for the same reason. When spring returns and starts to move into summer, you should recommission these as well as the air-pumps to increase oxygen levels in the pond as the water temperature starts to rise.
- Predominately a task for pump-fed systems, but one that may also be required in larger ponds where not all the sediment is pulled by the drains, is vacuuming the pond. The build-up of sediment on the bottom of the pond will not only harbour disease, but it will use up oxygen and create poor water quality. It is best removed.
- Check and clean filters as required. This is an ongoing job throughout the year and at least one major clean should carried out in the course of the year. Systems in which the discharge facilities from the filter chambers are limited will probably require more frequent cleaning; those that are discharged on a regular basis may only require looking at periodically. When cleaning any biological filter media, it is vital that pond water is used, as the chlorine in tapwater will destroy the beneficial bacteria found on the filter media.

Half-yearly tasks

- If running UV units, you should change the bulbs ideally every six months, or at the very least once a year. Although UV tubes will function for more than six months, after this time they do not operate at 100 per cent efficiency and so their ability to reduce green water is degraded.

Yearly tasks

- Any additional equipment associated with the pond, such as ozone units, requires yearly maintenance – the probe for the redox meter will need replacing, for instance. Make a note

Pond sludge is discharged through this tube.

Extension tubes allow the suction nozzle to reach into deep water.

Above: Pond vacuum cleaners such as this can be used to remove sludge and debris from the pond floor or filter.

of any special service requirements and ensure they are carried out at least once a year, or as otherwise suggested by the manufacturer.

- Make any major changes and modifications. Rather then disturbing your koi continually throughout the year, try to complete all major tasks at around the same time to limit stress to the fish. It is not ideal to move or disturb your koi, but if it has to be done, once is better than three or four times in the course of a single year.

By carrying out these tasks regularly, your pond will become a healthy and more stable environment for your koi and be more enjoyable for you. Routine maintenance ensures the pond and filters work effectively, leading to consistently good water quality. A stable environment promotes healthy koi for you to appreciate.

WHAT AND HOW TO FEED YOUR KOI

What your koi are fed is an important decision, as ultimately it will have an effect on their overall well-being. It is vital that the correct foods are fed at the right times of year to avoid problems. Koi are omnivorous which means they will eat both meat and plant matter, and in most instances their nutrition comes in the form of a processed fish food which will be either in stick, pellet, or paste form. Floating sticks are a popular choice but they can prove expensive as each stick contains a lot of air. Pellets are a better food source as each one contains far more substance than a comparable food stick. Paste food is also a good choice, but many people

are put off because it has to be prepared each day. Having said this, it is ideal as a treat or for feeding large individual fish as extra-large pellets can be formed from the paste. It is also important to ensure that the correct size of pellet is chosen for the size of fish in the pond. Many keepers now recommend that even large koi are better fed a medium-size pellet (6mm diameter) rather than large ones of 8mm and above.

Also consider feeding a certain amount of sinking food, as this not only encourages your koi to feed at different levels of the pond allowing more of the koi to get some food, but it also prevents damage from occurring when the koi

Above: Lettuce is a valuable source of vitamin C and other nutrients. Start by offering shredded leaves; in a short time koi will enjoy chasing a whole lettuce around the pond and tearing pieces off it.

1: To make paste food, measure out the required amount of powder.

2: Add water to the powder as directed by the mixing instructions.

3: Mix water with the paste until a dough-like consistency is achieved.

4: Roll and shape the paste in your hands to form large pellets of food.

5: Roll pellets to suit the size of koi in your pond. Remember that it will sink, which will encourage your fish to feed at different water levels.

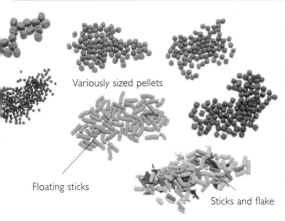

Variously sized pellets

Floating sticks

Sticks and flake

Above: Koi food is available in many different shapes and sizes. The type of food you give will partly depend on what time of year it is and how big your fish are.

come to the surface in a feeding frenzy and bump into one another. There are also health benefits to using sinking food as excess air is not taken in (see pages 118-9 for more information on swimbladder disorders). Whatever form of koi food you choose, it will be made up of a number of constituents. Protein is important for growth, tissue repair and reproduction. Protein cannot be stored in the koi body, so it is either digested and utilized or broken down to ammonia and excreted. It is important to give koi the appropriate amount of food. Offer high-protein diets when the weather is warm, as these will be used effectively for growth, tissue repair and reproduction. Fatty acids are also important, because koi use these as the main source of energy and they are essential for cell membranes and the immune system. Other main constituents of koi food include carbohydrates which are used for energy, vitamins which are essential for the well-being of your koi, and minerals like calcium, which is vital for bone structure, and sodium and potassium, which help to maintain the nervous system. The final constituent of any good quality food will be trace elements which include iron, manganese, zinc, iodine and others.

Right: Propolis can be added to your standard koi food by adding a measured amount to a weighed quantity of food.

Far right: Having added the propolis to the food, it should be thoroughly mixed, then allowed to dry before feeding to the fish.

What and When To Feed?

A koi's metabolic rate has a dramatic effect on its ability to process food. As temperature has a direct link to metabolism, the temperature of your pond will ultimately dictate what and how much your koi should be fed. If your pond is not heated, you will need to change your feed to a wheatgerm-based diet. This is generally lower in protein, and so is more easily processed by the fish. As a general guide, once water temperatures fall below 13°C (55°F) wheatgerm should be fed until the water temperature falls below 9°C (48°F) at which point feeding should be stopped altogether. Feeding should not be resumed until the water temperature has risen above 9°C (48°F), and wheatgerm should again be the choice of food. Once water temperatures are stable above 13°C (55°F), it is possible to opt for a standard staple diet, which will be suitable to

feed to your koi for the rest of the year, although in the summer months when water temperatures reach 18°C (64°F) and above, you may wish to supplement this with a high-protein growth food, or specialist foods to enhance colour.

If the pond is heated and maintained at a minimum temperature of 14-15°C (57-59°F) over the winter, you can continue to feed throughout the year. However, it is still advisable to give your koi a month or two on wheatgerm, which is readily digestible. During the summer, koi increase their fat reserves and use these to enable them to overwinter successfully. Day length is a significant factor in controlling fish metabolism and more important than heat. Along with proprietary koi foods, you may wish to supplement your koi's diet with the occasional treat. Oranges, brown bread, lettuce, prawns and other shellfish are all permissible on an occasional basis.

Oranges A good source of vitamin C which helps to reduce stress and boosts the immune system.

Garlic Garlic is a real treat for koi which are attracted to any food coated in it. It can also be used to entice them to feed by hand.

Lettuce Some koi keepers feed whole lettuce to their koi which readily shred and eat this treat which is a rich source of valuable nutrients.

Brown bread This is a good source of wheatgerm and vitamins, but feed sparingly as it is rich in carbohydrate.

Prawns Koi thoroughly enjoy prawns, a good source of protein. Feed in the summer when water temperatures are higher.

Bloodworms It is best to use frozen bloodworms as live food carries the risk of introducing disease into your pond. Offer bloodworms in the summer.

Above: A good diet combined with good system husbandry and water quality will help to keep your koi healthy.

Some people also decide to add live food, such as bloodworm or tubifex, to their koi's diet. However, we consider that it is best to avoid the use of live food because of the associated risks of introducing disease to the pond. The better option is the freeze-dried alternative, as this is irradiated before packing and so is disease-free. No matter what you are feeding, you must ensure that the correct amount of food is given, as overfeeding will result in water pollution, while underfeeding will cause your koi to become malnourished. As a guide, during the summer months when water temperatures are above 18°C (64°F) your koi can consume up to two per cent of their body weight a day (very young koi and fry may eat five times this amount!), so if you have 100kg of koi in your pond, you would need to feed 2kg of food per day. This should not be done in one feeding session; it is better to feed your koi small amounts throughout the day. If this is difficult for you, automatic feeding equipment may be a better option. As a general guide, do not feed more food than can be finished completely in five minutes.

Proper storage of dry food is important, especially once the seal is broken. Food should be stored under dry and cool conditions in re-sealable containers so as to minimize the degradation of vitamins (notably vitamin C) and other unstable ingredients. It is also important to realize that it is not just the quantity of proteins etc. that makes a good koi food, but also the quality of the ingredients. So do not be tempted by cheap unlabelled foods – only buy from reputable fish food manufacturers. If a well formulated dry food is given, there should be no need to mix in extra vitamins and trace elements.

Additives

Many koi keepers now mix additives with their chosen brand of food. These include propolis for its health and immune system benefits, vitamins and trace elements, or spirulina to enhance red pigmentation. Whatever additive is used, it is generally necessary to hand-mix this with the feed. To maintain freshness this should be done either daily or every other day using only the volume of food that will be fed. Additives allow you to administer ingredients such as probiotics with food, but it should be remembered that manufactured formulation feeds are balanced diets, with adequate vitamin and mineral content.

SPECIALIST POND EQUIPMENT

In addition to the basic essentials needed to maintain a healthy koi pond, you may want to use additional equipment to help maintain a better environment for your koi, and so reduce the chances of disease.

Pond Heating

Many now consider this a must, rather than a specialist piece of equipment. However, there are still many koi ponds which are not heated. Heating your pond can be achieved by the use of an electric in-line heater or by a heat exchanger, which will either be in-line or of a design which is actually submerged into the pond or filter, and powered by natural gas, oil, or propane gas. Pond

Heat exchanger for gas-fired boiler

A heat exchanger supplied with hot water from a boiler is a very cost-effective way of heating pond water. The exchanger is best mounted vertically.

Water inlet from filter system.

Warmed water returns to the pond.

Exchanger

Pump

From pond

Boiler

Thermostat

Left: Hot water from a boiler flows through the centre of the unit. Fins extending from the central pipe act like radiators to heat the water flowing round them.

Above: Most koi in Japan are kept in heated ponds; they will find living in unheated water an unusual experience.

heating means that the pond is kept warm all year round, and not simply kept from freezing in winter. There are numerous health benefits to this, namely:

1 Nearly all koi bred in Japan are kept in heated water and never experience temperatures lower than 15-16°C (59-61°F). The temperatures which they would experience in an unheated pond in other countries are alien to them and thus present a possible health risk.

2 In an unheated pond large temperature fluctuation may occur, which causes stress to koi. By heating you have complete control over the temperature and can provide a more stable environment for the fish.

3 In an unheated pond at temperatures below 10°C (50°F) the specific immune system of koi is ineffective,

but the non-specific immune system remains active. However, both bacteria and parasites are also inhibited by low temperatures. Medications are actually less toxic at lower temperatures and may be used.

4 In unheated ponds koi stop growing in very cold periods, and will not feed in the coldest months when water temperature drops below 9°C (48°F). Maintaining a higher temperature allows for all-year-round feeding which improves growth rates. You will also enjoy your fish more because they are not dormant for three or four months of the year.

If you do heat your pond, maintain it at the following temperatures throughout the year: December, January, February = 15-16°C (59-61°F), March and November = 18°C (64°F), the rest of the year = 22-24°C (72-75°F). Changes in pond temperature should be made gradually at a rate of only a degree or two every few days. Many people are worried by the cost of running a pond heater. However, if you have a collection of fair-sized koi (40-50cm/16-20in and above) and one dies during the winter, the cost of replacing it would be far greater than the cost of heating the pond for the winter and possibly for the whole year!

In-line electric water heater

This is the control unit of the water heater; it houses the timer clock, the thermostat and a digital readout which displays the water temperature.

This probe in the pond monitors water temperature and relays the reading to the display and thermostat in the control unit.

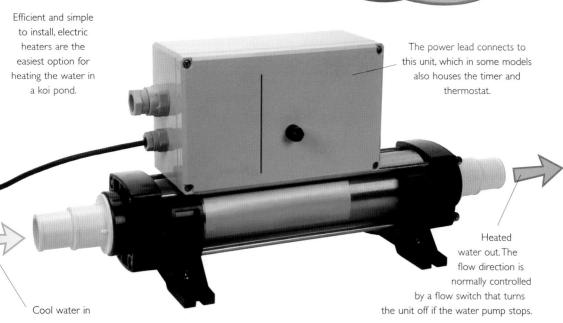

Efficient and simple to install, electric heaters are the easiest option for heating the water in a koi pond.

The power lead connects to this unit, which in some models also houses the timer and thermostat.

Heated water out. The flow direction is normally controlled by a flow switch that turns the unit off if the water pump stops.

Cool water in

Ozone Units

Ozone generators are commonly used by marine fishkeepers, but only recently have units become available for pond use. They are still considered very specialist items and this is reflected in their generally high price. Ozone consists of an oxygen (O_2) molecule with an added oxygen atom, creating ozone (O_3). Ozone is an unstable molecule, which makes it a powerful oxidizing agent, effective against viruses, bacteria and protozoa, either directly as the third oxygen atom detaches on contact with any organic particle, or indirectly, as ozone decomposes to release free radicals, atoms with unpaired electrons that damage nucleic acids of the cells. Ozonolysis occurs when the third oxygen atom attaches to molecules with double bonds, to produce two simple molecules with different properties and chemical characteristics. To reduce foam build-up, it is a good idea to operate a protein skimmer if running an ozone unit. Ozone can be quite harmful if used in a confined space, and may cause headaches and sickness if breathed in. The presence of ozone can easily be smelt, but be aware that ozone can be harmful to humans at levels well below those detected by the nose. To avoid any problems it is vital that a unit be selected which either removes any excess ozone before it is released into the air, or else install the unit in a fully ventilated location. There are associated risks with the incorrect use of ozone, so it is vital that usage levels are controlled. This is generally done by using a redox potential meter which ensures a correct ozone dosage.

Although ozone is an excellent disinfectant, ozonizers will not create a sterile system because a pond is an open, exposed system. Certain protozoa, such as *Trichodina* species, are free living and associated with the filtration system. Also, koi may be carriers of certain diseases and so it impossible to achieve a completely disease-free status. Ozone will reduce the bacterial load and in the event of a disease outbreak will lower the numbers of pathogens. If used correctly, an ozone generator is an excellent addition to any pond, as it will reduce the factors responsible for many health problems and improve water quality and clarity. It is advisable to run ozone in conjunction with your existing settlement and biological filter and treat it as an addition, rather then relying solely upon a settlement chamber, such as a vortex, and an ozone unit.

Ozone unit

UV unit burns off excess ozone in water.

Inlet for ozone mix with water.

Valve to control water flow.

Gravity return to pond.

Probe from redox meter inserted here; it measures ozone levels in the water.

Water pumped in from last filter chamber.

Ozone generator

Discharge to waste

Ozone gas reacts with pond water in this chamber.

Ultraviolet (UV) Clarifiers and Sterilizers

The UV systems used in most ponds are clarifiers. They help to prevent green water by killing the algae responsible for it as the water passes through the unit. It is possible, however, to use UV light to sterilize the water, and this is achieved by the use of UV sterilizers. These work by exposing the water from the pond to the UV light at much higher concentration, and this is achieved by only having a small distance between the bulb and the outside of the casing through which the water flows. Consequently, a much higher amount of UV light is required to sterilize a body of water effectively and this unfortunately makes UV sterilization of koi ponds impractical for most

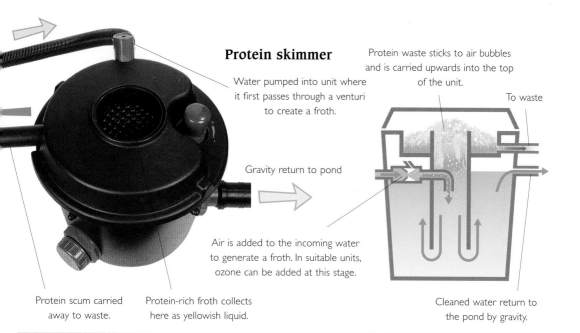

Protein skimmer

Water pumped into unit where it first passes through a venturi to create a froth.

Protein waste sticks to air bubbles and is carried upwards into the top of the unit.

To waste

Gravity return to pond

Air is added to the incoming water to generate a froth. In suitable units, ozone can be added at this stage.

Protein scum carried away to waste.

Protein-rich froth collects here as yellowish liquid.

Cleaned water return to the pond by gravity.

people because of the volume of water which they contain and the amount of UV light thus required. However, it is a viable proposition for smaller quarantine tanks. If you decide to buy such a unit, you will need to discuss your exact needs with your koi dealer or an equipment manufacturer to ensure that a UV sterilizer is purchased and not a standard UV clarifier as is normally the case.

UV clarifier

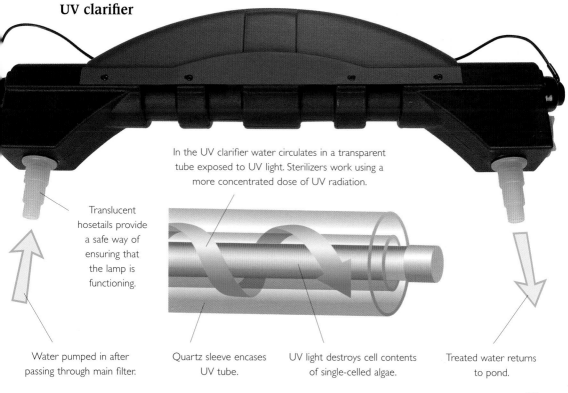

In the UV clarifier water circulates in a transparent tube exposed to UV light. Sterilizers work using a more concentrated dose of UV radiation.

Translucent hosetails provide a safe way of ensuring that the lamp is functioning.

Water pumped in after passing through main filter.

Quartz sleeve encases UV tube.

UV light destroys cell contents of single-celled algae.

Treated water returns to pond.

ESSENTIAL SKILLS

Having ruled out poor water quality as the cause of any problems, the next step in determining why your koi are not behaving as you would expect is to start looking for signs of disease and, if found, to identify exactly which disease it is. Too often people see their koi hanging in the water, flicking against the bottom, or gasping for air around airstones in the pond, and simply dose the water with an "anti-everything" treatment. Although this is better than doing nothing, it can create its own problems for many diseases require specific treatment and this scattergun approach generally will have no positive effect. Before any medications are added to a pond, it makes sense to know exactly what you are treating.

This section of book first looks at the anatomy of koi to give you an insight into how different parts of the fish's body function and what their physiological role is. This helps you to understand what effect a specific disease will have on the overall well-being of your koi, depending upon which part of the body is under attack. It will help you to appreciate the fish's external anatomy and where the internal organs are positioned and what their function is. This knowledge all contributes to your understanding of koi healthcare.

Next netting and inspecting koi is described. This is an important part of the process of disease identification; if it is done incorrectly, it can result in physical damage occurring which in turn may lead to other secondary infections. When netting koi you must use the correct equipment. If you do not own a viewing bowl, pan net and sock net already, they should be considered a vital addition to your koi first-aid kit.

Regular correct inspection of your koi is sensible as it often allows you to identify potential problems before they become serious. If you do spot a problem, it is a good idea to take a skin scrape or swab and these two topics are covered next. If taking a skin scrape, it is best to have your own microscope to examine the sample, and you will find a useful guide to buying and using a microscope in this section. A microscope is another piece of equipment that the serious koi keeper should not be without. Both skin scrapes

and swabs are simple procedures which can be carried out at home, and a detailed explanation of what is involved is given. However, the analysis of a swab is an involved process and it has to be sent away for testing to a suitable laboratory.

Finally in this section the procedures involved in transporting koi are examined. You may want to take your koi to a show, or need to get one to your local vet or koi dealer for examination, and it is vital that this is done correctly. Transporting koi incorrectly can be highly stressful and may cause opportunist diseases, such as those caused by parasites, to become a problem. One of the hardest things to guard against, but one which can cause the greatest stress to your koi, is temperature change. It is vital to try to regularize the temperature between the water into which the koi will be put and that in the transport bag. If your koi is not going into a new system, but is simply being taken for veterinary examination and then being returned, you must ensure that the temperature in the bag is not allowed to fall or rise too much. Tips on how to achieve this are given in this section.

If all these procedures are adopted and used on a regular basis as part of the routine maintenance of your pond, you will be far more likely to identify any problems early. You will also have absorbed the information needed to use section 3 of the book effectively. You will know which condition you are treating; as a result the correct course of treatment can be administered, thereby considerably increasing the chances of a rapid and full recovery.

Take Care!
While experienced hobbyists may feel comfortable carrying out the techniques described in this section, always consider the well-being of your fish first. If you are not an experienced fishkeeper, or are in doubt about any of the procedures, always consult a vet or koi health specialist first.

Right: It is important to learn the correct ways of handling koi to avoid causing them undue stress.

KOI ANATOMY

It is useful to have a basic understanding of the anatomy of your koi. Not only does this help in understanding how and why certain diseases cause the problems they do, but knowledge of what each part of the body does also helps in the identification and treatment of particular ailments which attack specific parts of the body.

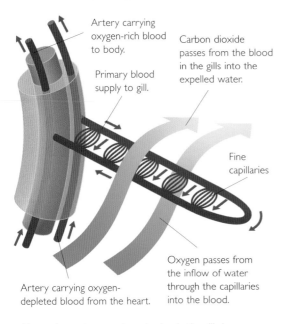

Artery carrying oxygen-rich blood to body.

Carbon dioxide passes from the blood in the gills into the expelled water.

Primary blood supply to gill.

Fine capillaries

Artery carrying oxygen-depleted blood from the heart.

Oxygen passes from the inflow of water through the capillaries into the blood.

Above: A countercurrent mechanism in the gills is very efficient and ensures that the blood absorbs the maximum amount of oxygen from the water.

The Skin, Scales and Lateral Line

A koi's skin is covered by a layer of mucus that not only offers protection against disease, but also helps streamline the fish to aid locomotion. This layer of mucus is continually being refreshed as the old mucus sloughs off into the water. Below the mucus lies the top layer of skin known as the epidermis. This is extremely thin and in fact lies above the scales. The epidermis is responsible for mucus production and assists wound repair by multiplying to cover areas of damage. Under the epidermis is the dermis, where the scales are formed. Scales are flexible, regularly arranged and overlapping bony plates that help to protect the body surface and also help to streamline the koi, reducing friction with the water as it swims. The scales tend to be uniform in shape with the exception of those which form the lateral line. These have a tiny opening in them which appears

to the naked eye as a line running along the length of the koi. The lateral line is a sensory organ used to detect vibrations in the water. Hair cells transmit any vibration via nerve fibres to the spinal cord which sends a signal to the brain. The dermis also contains fibrous connective tissue for elasticity and strength, blood vessels, nerves and chromatophores, and pigment cells responsible for coloration.

The Gills

The gills are made up of four bony arches which are the main support for the lamellae (or gill filaments), which lie in a V formation. These primary lamellae have a large surface area and contain many blood capillaries which allow oxygen to diffuse from the water into the blood. The gills are protected from damage by a bony cover known as the operculum. It is the flow of water over the gills that enables koi to extract

Freshwater osmoregulation

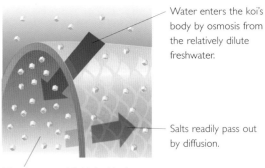

Water enters the koi's body by osmosis from the relatively dilute freshwater.

Salts readily pass out by diffusion.

The tissues and fluids inside the koi's body contain more salts than the surrounding water.

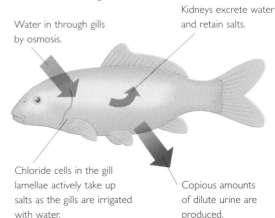

Water in through gills by osmosis.

Kidneys excrete water and retain salts.

Chloride cells in the gill lamellae actively take up salts as the gills are irrigated with water.

Copious amounts of dilute urine are produced.

Skin, scales and fins

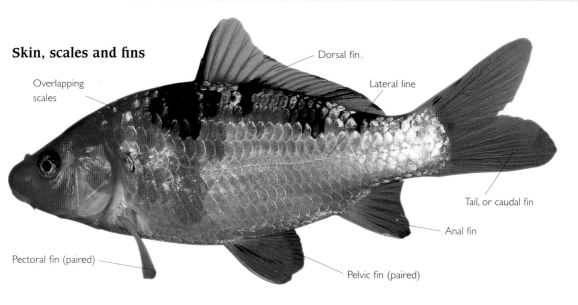

Overlapping scales

Dorsal fin

Lateral line

Tail, or caudal fin

Anal fin

Pectoral fin (paired)

Pelvic fin (paired)

oxygen from the water for their survival. The process by which water is passed over the gills is known as the buccal pump; and it has two parts. First comes the water intake which is achieved by the fish opening its mouth while holding its operculum shut. Water collects in the mouth and the area under the mouth (known as the buccal cavity) which is now lowered. Then comes water expulsion when the fish closes its mouth while lifting its buccal cavity; this has the effect of forcing oxygen-rich water over the gills and out of the now open operculum. To optimize the diffusion of oxygen into the blood that flows through the membranes in the gills a countercurrent mechanism is employed. This means that the water flowing across the gills travels in the opposite direction to the blood. As the water passes over the lamella in the opposite direction to the blood flow, it always has a higher dissolved oxygen concentration than the blood cells with which it is coming into contact. This optimizes the amount of oxygen that the red blood cells can absorb by diffusion. The main advantage of the process is that it allows for diffusion to take place along the whole length of the gill lamellae.

Osmoregulation (Gills, Kidneys, Skin)

This is the process by which koi control the balance between fluid and salt levels within their cells. As koi are freshwater fish their bodies have a higher salt concentration than the water around them, and consequently the salts in their cells

want to diffuse out into the surrounding water. Similarly, water enters the koi mainly through the gills and the skin, both of which are partially permeable. To help maintain the correct salt level within a fish, paired kidneys actually extract the required amount of salt, while producing urine as a waste product which is discharged via the vent. In addition koi control their gill movements so that their gills are exposed to the pond water for only a limited period of time, so reducing the levels of salts that diffuse out. Chloride cells located on the gills also help matters by taking up salts from the water that passes over them.

Swimbladder

This is an internal organ consisting of two chambers, and its purpose is to control buoyancy as well as to transmit sounds to the inner ear via a number of small bones known as the Weberian ossicles. The swimbladder in a koi is a gas-filled sac, and the level of inflation is generally controlled by blood vessels lying around it. Because the swimbladder is connected to the gut, koi can increase the level of gas within it simply by gulping air from the pond's surface.

Digestive Organs (Oesophagus, Liver and Intestine)

Koi do not have teeth – instead they have paired pharyngeal bones which are used to crush the food as it passes through the muscular oesophagus into the intestine where it is digested and absorbed into the blood and carried to the

Internal anatomy

The gills are the site of gas diffusion. Carbon dioxide is released into the water and oxygen is collected by red blood cells.

Sound waves are detected by the swimbladder and amplified by a series of modified backbones, linking them to the inner ear, enabling the fish to hear.

The posterior kidney conserves salts in the body and produces vast amounts of very dilute urine to remove water and maintain the correct osmotic balance.

The swimbladder is a gas-filled buoyancy organ that allows the fish to remain at any depth in the water with the minimum expenditure of energy.

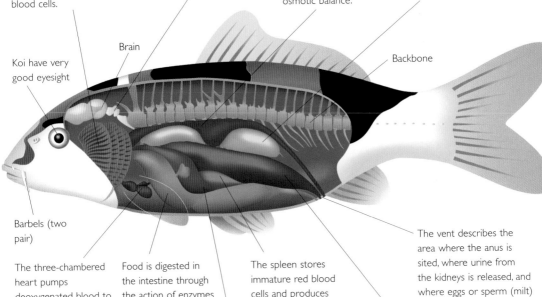

Brain

Koi have very good eyesight

Backbone

Barbels (two pair)

The three-chambered heart pumps deoxygenated blood to the gills. The muscular section of the heart (the ventricle) receives oxygen-rich blood from the coronary artery.

Food is digested in the intestine through the action of enzymes and absorbed by the blood supply.

The spleen stores immature red blood cells and produces cells of the immune system.

The vent describes the area where the anus is sited, where urine from the kidneys is released, and where eggs or sperm (milt) are shed into the water.

The liver stores or distributes digested food to the tissues, and breaks down unwanted proteins into waste.

The gonads are situated on either side of the body. The ovaries produce eggs while the testes produce sperm.

liver. Any food not broken down by the digestive enzymes will be voided at the vent as waste. Koi do not have stomachs – this is why the food goes straight from the muscular oesophagus into the intestine. The liver in koi is large and usually deep red in colour, although it may become discoloured due to the formation of hemosiderin, a by-product from metabolism of old red blood cells. It deals with the distribution of nutrients, as well as with the breakdown of unwanted proteins which are turned into waste ammonia. The liver will also remove any toxins or poisons, as well as breaking down old red blood cells into bile.

Sight (Eyes)
Koi, like many other fish species, have excellent eyesight. Many experts think that they actually see things in a very similar way to humans, and are able to make out shapes and colours.

Fins
Koi have a number of fins. Some, like the pectoral fins, are paired while others are single, like the dorsal. These fins have various uses including providing propulsion, steering, and helping to maintain stability within the water.

The Heart and Circulation System
A koi's heart contains three chambers: the sinus venosus, which receives blood from the veins, the atrium, which pumps the blood into the ventricle, the main pumping chamber sending blood through the aorta and arteries. Blood circulates from the heart to the gills, where it takes on

Above: When a koi comes to the surface to feed, it uses its fins to propel itself and to maintain its position in the water.

oxygen, and then to tissue matter. At this point it has a relatively high pressure. However, as this blood flows to the extremities of the body it becomes slower and the oxygen level decreases. 50 per cent of the blood is made up of blood cells. Most are red cells that carry oxygen around the body, and the remainder are white blood cells which are important to the koi's immune system. The rest of the blood consists of plasma which contains water, salts, glucose, plus any waste which is being transported around the body.

Reproductive Organs (Gonads)

These are located internally and are found on either side of the body. Female koi have ovaries, responsible for the production of eggs. In some instances they can be very large (see pages 88-91 on egg retention). Testes are found in male koi and are responsible for the production of milt (sperm). Both eggs and milt are released by a koi through its vent into the water.

The Immune System

The immune system comprises many elements. Firstly, the mucus is continually sloughed as the koi swims and contains elements of the non-specific immune system, special chemicals and cells that can eradicate foreign bodies. Two skin layers beneath the mucus serve as physical barriers: the outer epidermis and the inner dermis within which the scales are rooted. Many infections take hold by entering through breaches in the skin. The digestive tract contains a vast bacterial flora, that aids the breakdown of food and helps combat any pathogens that are ingested.

The immune system is made up of the non-specific and specific immune system. The non-specific immune system remains functional irrespective of temperature, with a number of cells and cytotoxic cells that will engulf or kill any foreign bodies encountered both internally and within the body surface of the koi. The specific immune system is temperature-sensitive, as it is energetically expensive to maintain. It relies on specialist antibody cells that recognize particular pathogens and trigger a cascade of immune cells to fight an infection. The ability of the specific immune system to recognize target pathogens is the basis on which vaccinations work. The vaccine presents either dead or modified pathogens to the immune system, which is then triggered into action if that pathogen infects the koi.

NETTING AND HANDLING YOUR KOI

If you believe that your koi are suffering from any of the health problems which are discussed later in this book, you will probably need to take one or more fish out of the pond for closer inspection. This could be in order to take a skin scrape, a mucus swab, to move the fish into quarantine or simply to take a closer look at the fish before deciding on the next course of action. When moving or examining koi, you must do so in such a way as to cause the minimum stress or injury to the fish. Bad netting and handling can itself be a large stress factor which could lead to future health problems, and many physical injuries are caused by careless netting.

A sock net

A pan net

Catching Your Koi

Before any attempt is made to catch a fish you must have the right equipment:
- a pan net suitable to accommodate the largest fish in the pond.
- a viewing bowl.
- a sock net.

A pan net is very shallow, as this reduces the risk of the fish becoming entangled in the netting. In fact this net should only be used as a guide and under no circumstances should you ever lift the net out of the water with the fish in it. You are basically trying to use it to shepherd the desired fish into a viewing bowl. Ideally get a helper at this point to submerge one side of the bowl so that the fish can simply swim from the net into the bowl. If you have to work on your own, float the viewing bowl and half fill it with water. Once the fish is caught, bring it to the surface and lift one side of the net so that it hooks over the lip of the

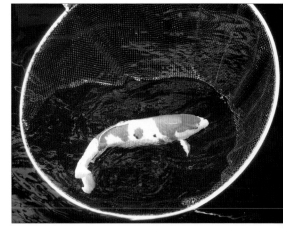

1: Catch the koi to be inspected in a suitably sized pan net.

2: Guide the koi to the bowl with the net, then use the edge of the net to tip one side of the bowl below water level.

3: Gently tip the net to transfer the koi from net to bowl.

bowl. Then force one side of the bowl down into the water and simultaneously tip the fish from the net to the bowl. The viewing bowl used should be like the net: i.e. its diameter must be wide enough to accommodate your largest koi. It is advisable to opt for a floating bowl or else you may find you are spending more time lifting the bowl from the bottom of the pond than you are catching koi. When actually trying to catch a fish there are a few simple rules which should be followed. These will make it easier for you to catch the koi and reduce the risk of damage being caused to it.

1 Turn off all air pumps and water features, this will let you see what you are doing.

2 Move the net so that its edge slices through the water. This reduces drag and allows you to move the net much more quickly.

3 When bringing a captured fish to the surface avoid any underwater returns. The pressure of the water returning to the pond from these could force the koi into the net and cause it physical damage.

4 If you are trying to catch small koi or the pond is very large, you may find it hard to catch the fish with one net. Rather than chasing around your pond and risking accidentally bumping the fish with the net, try one of two tactics. Either ask someone to help you and give them another net so that they can keep the fish at one end while you try to catch it. If this is not possible, lower the water level to reduce the area in which the fish can swim.

5 If the pond is very large, netting a particular koi may prove too stressful to both yourself and the fish. In such a case it may be advisable to get

1: If you have a helper, ask him to hold the bowl as shown.

2: Move the net towards the submerged edge of the bowl.

3: Gently tip the net so that the koi swims into the bowl.

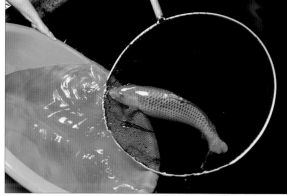

4: To prevent jumping, keep the water level quite low.

1: Pull the sock net through the water to make it wet.

2: Coax the koi head-first into the sock net.

3: Hold each end of the net and lift it out of the water.

4: Pull the sock net through the water to release the koi.

a custom-made seine net for your pond which allows all the koi to be caught easily without undue stress. Then the ones which need to be inspected can simply be removed and placed into a viewing bowl, before the others are allowed to return to the pond.

Once you have a fish in the viewing bowl, you may wish to move it or to inspect it more closely. There are four ways in which you can move fish without causing unnecessary stress and damage. They are:

1 Sock net

This is a net which is shaped very much like a table tennis bat with a long, open-ended, sock-like net trailing behind it. They are available in numerous sizes, and it is important to ensure that the one you are using has a large enough diameter for the fish in question, and that it will hold the full length of the fish in the sock. Before attempting to move the fish make sure that the whole of the sock is wet. Then hold the sock net in one hand under the water in the bowl and use your other hand to persuade the fish gently into the sock head first. Once fully in the sock, use one hand to hold the end of the sock while the other hand holds the handle. The sock net can then be lifted out of the water. Make sure that the net is kept as straight as possible. To release the fish from the sock simply immerse it in the water and let go of the end of the sock. Then lift the sock from the water by the handle, and if done correctly the fish will swim out.

2 Plastic bag

If a sock net is not available, a plastic bag can be used in much the same way. The only difference is that as the bag is sealed at one end, you will have to tip the fish from the bag.

3 Moving the bowl

If neither a plastic bag or sock net are available you can quite simply carry the viewing bowl to where you wish to move the fish. This may sound simple, but normally the bowl is heavy due to the volume of water it holds, and it may need two or more people to lift it. There is also a risk of the fish jumping in the bowl, and damaging itself. This method of moving fish is better suited to smaller koi less than 30cm (12in) in length.

4 Carrying the fish

This method should not be attempted unless you are confident at handling fish, and really it should not

1: Put the plastic bag in the bowl, half-fill with water, then coax the koi into it.

2: Ensure enough water to cover koi.

3: Hold the top and one of the bottom corners of the bag.

4: You can now lift the bag and move the koi as desired.

be considered if a sock net or plastic bag is available. The fish is quite simply picked up from the water and carried to its new location. When carrying a fish, one hand is normally placed under the pectoral fins while the other goes just behind the anal fin.

1: To move a fish by hand, position your hands as shown.

2: Ensure that your hands are comfortable, then lift the koi from the water. If the koi starts to move, do not tense – simply let your hands move with the fish.

The hard part is to know just how tight to hold the fish, and not to be nervous as this causes the fish to struggle. Really you should not grasp the fish but just let it rest in your hands. Some people say it is best not to look at the fish as this causes you to worry too much about what it is going to do. If the fish starts to struggle, you must stay relaxed and let your hands go with the movement of the fish. Once the fish relaxes, you can generally carry it without problems. The risk with hand-moving fish is that if one does panic you may drop it, and cause considerable damage and stress. This is more likely to happen with male fish which are more aggressive or Doitsu/leather varieties (no scales or a single line of scales) as they are more slippery.

1: To tip a fish, hold it firmly against the side of the bowl.

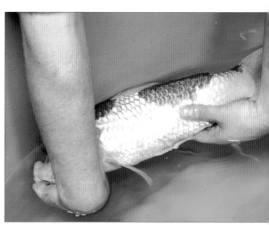

2: Rotate the koi against the side until it is upside down.

3: It can be held against the bowl ready for inspection.

Once you have your koi bowled, you can inspect it for damage, parasites or take samples from it while it remains in the bowl. But if you want to inspect the underside of your koi for any problems, you have various options.

Hold the fish out of water
Follow the steps for carrying the fish, and lift it out of the water while someone else inspects underneath it.

Tip the fish in the bowl
If unsure about lifting the fish from the water, you can turn the fish upside-down in the bowl. You should push the fish against the side of the bowl – hold the fish quite firmly while doing this. Once you are happy with your grip on the fish, turn it over in your hands, so that it is upside-down in the water, but still against the side of the bowl. After inspection simply loosen your grip and the fish will return to an upright position.

Hold the fish in a plastic bag
An alternative approach is simply to bag it in a clear plastic bag and hold this up so that you can view the underside of the fish.

Use sedating agent
If unsure about the methods suggested, you can sedate (anaesthetize) the fish (see pages 132-4) and inspect it while it is under sedation. However, if you wish to take a skin scrape or swab, it is best to do this without the use of sedation as this may alter the results.

Spinning the fish
If you decide to use one of the handling methods, you may wish to employ a tactic called "spinning". You simply spin the fish gently with your hands while it is in the bowl for a minute or two. This causes the fish to become disoriented which in turn makes it easier to handle without the need for a sedating agent.

Do remember that if you have to net, inspect or move koi, if you try to do it in haste, with the wrong equipment or without correct planning, you could make a problem a lot worse or even create new ones. If you are ever unsure about the correct procedure, be sure to seek advice from your local koi specialist.

1: Using your hands, gently spin the koi with a circular motion.

2: Continue doing this, working in the same direction.

3: After a minute or two the koi should be quite calm.

USING A MICROSCOPE

A microscope is an essential piece of equipment for the correct identification of diseases. It allows koi keepers to identify parasite infections. Many koi keepers cite the price of a good quality microscope as the reason they do not own one. But this is false economy. If you can identify parasites and treat them early, you can stop the spread of an infection before it becomes a real concern and leads to potential

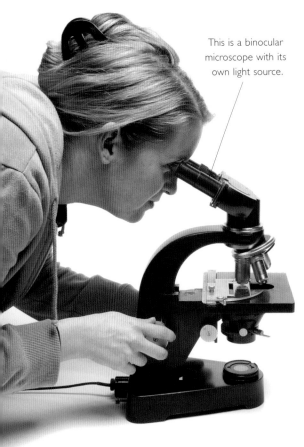

This is a binocular microscope with its own light source.

Above: A microscope is an essential tool for the koi keeper to aid in the identification of many diseases.

losses or physical damage to high quality koi. A good quality microscope is also a sound investment, and if the time comes when you no longer need it, as long as it has been well maintained you will easily be able to sell it again, and recoup some of your initial outlay. Various factors should be considered when deciding on the best unit for your budget and intended use.

Styles of Microscope

Firstly do you want a binocular (using two eyepieces) or a monocular (single eye-piece) microscope? This choice is largely influenced by budget and personal preference. Another factor which is often influenced by budget is the decision whether you require a microscope with its own light source or not. A microscope with its own light source generally needs mains power to light a bulb in the base of the unit. The advantage of this is that the microscope can be set up in one location and used at any time simply by turning on the power supply, while the intensity of the light can normally be controlled allowing greater flexibility. The other – generally cheaper – option is to buy a microscope which does not have its own light source. Instead it has a mirror located in the base, which is turned to reflect rays of light onto the subject being examined. The disadvantage of this is that you are dependent upon an external light source, and the intensity of this light may not be controllable which can affect the brightness of the image.

Lenses

There are two lenses which you need to consider when purchasing a microscope: the eyepiece lens and the objective lens. Most microscopes feature more than one objective lens, commonly three will be found each with different powers of magnification. These lenses sit on a revolving turret so that the magnification can easily be changed by simply turning a dial which will move to the next lens. A microscope mainly intended for parasite identification should be equipped with lenses of the following power – 10x, 20x and 40x. The eyepiece lens is what you actually look down, and its magnification will affect the overall magnification of the image. These are easily interchangeable and thus allow for different levels of magnification to be achieved. The typical magnification of the eyepiece lens for parasitic identification is normally 10x or 15x. If a 10x eyepiece lens is used with an objective lens of 40x, a magnification of 400x can be achieved (10 x 40 = 400). Do not worry that you are limiting yourself by purchasing a microscope at this magnification. As long as you buy a reputable make, you will be able to buy other more powerful lenses later on.

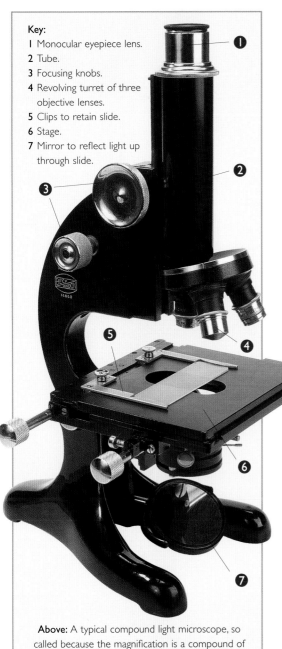

Key:

1 Monocular eyepiece lens.
2 Tube.
3 Focusing knobs.
4 Revolving turret of three objective lenses.
5 Clips to retain slide.
6 Stage.
7 Mirror to reflect light up through slide.

Above: A typical compound light microscope, so called because the magnification is a compound of the lens power in the eyepiece and objective lenses.

may either be fixed or move up and down. In some microscopes the stage moves up and down to bring the slide into focus, while others will work by moving the lens and eyepiece with a focus dial. The slide is held in place by a slide clamp. The sample on the slide is normally covered by a small square of thin glass called a cover slip.

Viewing The Slide

Once firmly in position the slide can be viewed. To prevent damage to the lens it is better to focus away from the slide. If you move the lens towards the slide you may over-focus and crash the lens onto it which can both break the slide, and more importantly damage or scratch the objective lens. There are typically two controls on a standard microscope for controlling the focus. The first of these should be used initially to allow an object to be found quickly and roughly focused. Then the second control allows for fine tuning of the focusing by moving either the tube or stage just a fraction each time the fine focus dial is turned.

Unless you know exactly what you are looking for and at what magnification it will be found, always start at the lowest magnification, then increase it once a full scan of the slide has been achieved. To ensure you see all of the slide, view it in a methodical manner making either left to right or front to back passes, before gently moving on to the next sector to be viewed. The slide is moved by use of controls situated underneath the stage. There may also be additional controls for finer movement which allows for exact positioning over a specific area.

Some microscopes have additional parts, such as a condenser, for focusing the light, or an area that can hold colour filters. They are not required for the simple identification of parasites, but they may have to be set up initially to allow the microscope to function correctly. You may find that once you have mastered the basic skills required to identify parasites that microscopy becomes a hobby in its own right. However, if you find the whole idea of using a microscope daunting, do not be afraid to speak to your local koi dealer and ask for guidance. Soon you will find that your microscope becomes a vital part of your koi healthcare arsenal.

Mounting A Slide

Having taken a slide for examination (see pages 44-5), it is essential that it is placed correctly on the microscope so that it can be viewed. The slide is placed is what is known as the stage, which

HOW TO TAKE A SKIN SCRAPE

If your koi are not behaving in their normal way, but all water quality parameters including the oxygen level are acceptable, it is worth taking a skin scrape. This involves scraping a sterile blunt instrument, or (for the more experienced) a slide, over a fish's skin to collect some of the mucus which covers its body. The mucus sample can then be examined under a microscope, and any parasites present within the pond that may be affecting their behaviour identified. Typical behavioural changes include koi hanging in the water, swimming with fins clamped to the sides, rapid gill movements, fish congregating in areas of enriched oxygen such as water returns and airstones, fish that are isolated and not mixing with other koi in the pond, loss of appetite, and flicking or rubbing against any surfaces in the pond as if the fish were trying to relieve an irritation. If you do not own a microscope to examine the skin scrape, a nearby koi dealer can help, but they must be able to view the scrape quickly, ideally within 30-60 minutes of it having been taken. And you must be sure that the dealer has the right facilities and sufficient expertise to be able to make an accurate diagnosis.

To take a skin scrape you will need the sterile blunt instrument and some slides and cover slips to hand. If you are taking a sample from a large koi, it also helps to have someone to help you, as you cannot sedate the koi before taking the scrape. A sedating agent can have an effect on the results from the scrape, and so large koi must be physically held while one is taken.

Taking A Skin Scrape

1 Place the koi in a viewing bowl. Then tip out some water so just enough remains to cover the koi. When dealing with small koi you may be able to hold the fish in one hand while taking the scrape with the other. For larger koi it is far easier if you have an assistant who can hold the koi against the side of the bowl while you take the scrape. It is sometimes necessary to apply quite a bit of pressure to the koi while holding it to stop it from jumping about and damaging itself. To calm the koi down you can employ a tactic known as spinning (see page 41).

2 Once the koi is firmly held, take the blunt instrument or slide and run it along the fish's body. Do this from head to tail in the direction of the scales. **Never** take a skin scrape in the other

1: Take a blunt instrument or slide, as shown here.

2: Gently run it along the body in the direction of the scales.

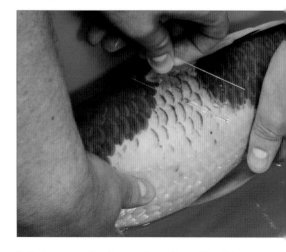

3: Body mucus will collect on the surface of the slide.

Right: Having taken your sample, it can now be examined. In order to prepare the slide, a cover slip has to be placed over the mucus. Then the slide can be positioned on the microscope ready for examination.

4: A collection of mucus will be deposited on the slide or blunt instrument used to take the scrape.

direction as this may result in scale damage. You may need to apply a little pressure while taking the scrape, and perhaps make more than one pass to get enough mucus for the sample. A skin scrape can be taken anywhere on the koi, but do take note of the symptoms which your koi are displaying. If, for example, it is gasping for air, take a scrape as close to the gills as you can. The amount of mucus required is quite small and as long as an area of the slide can be seen to have mucus on it, you will have enough to examine it successfully. If using a blunt instrument to take the scrape, the mucus should be transferred onto the slide once collected.

3 Once the scrape is taken you can put the koi back in the pond. Normally a scrape will not have any adverse side effect on the koi as long as it is done correctly.

4 You now prepare your slide for examination under the microscope by taking a cover slip, and

using this to move the sample of mucus to the centre of the slide. Put a drop of pond water (**not** tapwater) on the slide to dilute the mucus sample, then drop the cover slip into position and gently ease it onto the slide allowing the mucus to spread out underneath it. If you do not have any cover slips, you can use another slide and sandwich the mucus between the two pieces of glass.

5 Now examine the slide under your microscope. If nothing is found, it pays to take scrapes from other koi in your pond.

Only consider taking scrapes when signs of parasite infection become obvious. Remember parasites will always be present but in low numbers; it is only when koi become stressed that these numbers increase and a problem results. As you learn how to view and interpret a scrape, you will be able to spot any increase in parasite levels and hopefully identify and treat the problem before it gets out of control.

HOW TO TAKE A SWAB

If water quality is fine, and a skin scrape has not revealed an obvious reason why your koi are looking off-colour (perhaps ulcers, sores and areas of infection have appeared), it is worth taking a swab and sending it away for analysis to check if a bacterial infection is causing the problem. If koi are already showing areas of ulceration or if previously damaged or diseased areas do not heal, consider taking a swab straight away as these symptoms suggest a bacterial problem. Also test the water quality and take a skin scrape to check for parasites.

Unlike a skin scrape, which can be examined at home, a swab must be processed by a qualified laboratory. Your local koi specialist should be able to assist you by supplying the necessary equipment and either sending it away for you or advising you on doing this yourself. The results may also be sent back to your local dealer or vet, who can advise on the correct course of treatment. Try to take and send swabs to a laboratory at the start of the week, unless otherwise advised. If you take a swab on a Friday but the lab is closed over the weekend, work will not start on it until Monday at the earliest. By this time levels of bacteria on the swab may have decreased or increased and give a false reading.

Taking A Swab

1 Make sure you have all the necessary items, both for taking the swab and getting it in the post immediately, and have advised the laboratory it is being despatched. Generally you need the swab, a request form completed with your details and details of where and how the swab was taken, a pre-paid envelope, and a biohazard bag.

2 Now catch the koi and place it in an appropriate sized bowl. It is preferable to sedate the koi lightly, which allows it to be handled more easily and reduces stress associated with handling. Bacterial pathogens that cause fish disease are usually very sensitive and it may be difficult to isolate these using a swab as during transport they are readily outcompeted by the growth of opportunist species sampled at the same time.

3 It is far easier to carry out this step with two people. One of you should hold the koi against the side of the viewing bowl so that any areas

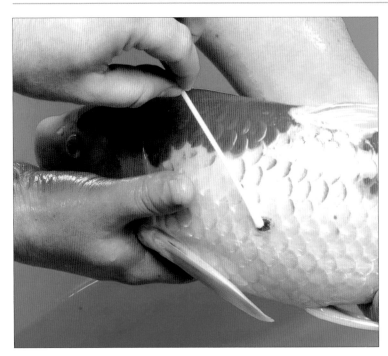
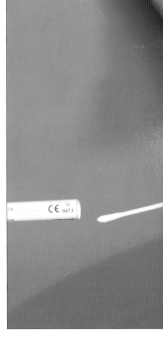

1: Remove the swab from its protective sterile packaging and rub it over the area to be tested.

2: As soon as you have taken the sample satisfactorily, place the swab back into its sterile protective casing.

which look infected are easily visible. Some pressure may be required to hold the koi still but this will cause less harm than if the koi is simply left to thrash about in the bowl. If the koi is particularly lively, a technique known as spinning can be employed to calm it down (see page 41). Once the koi is held firmly in position, the other person can take the swab from areas of ulceration or sites of physical damage on the fish as it is these areas which give the most accurate diagnosis of what is actually present.

4 Only open the packing when ready to take the swab; the actual swab should not be taken from its container until the koi is secure in the bowl. This will reduce the risk of any airborne bacteria getting onto the swab and giving false readings. A swab is basically like a large cotton-wool bud, and to take a sample it is simply rolled back and forth over the area to be tested.

5 Once the sample is obtained, immediately place the swab back into its container, again to prevent any contamination. Write any necessary details on the container for identification purposes.

6 The koi can now be released back into the pond – it should suffer no adverse effects.

7 Once labelled, and with all required paperwork completed, the container can be placed in the biohazard bag. This goes into the addressed envelope which should be posted immediately, unless you have arranged to return it to your koi specialist or vet instead.

Make sure you use the quickest postal service available – many laboratories will supply pre-paid envelopes. The quicker the swab gets to the lab, the sooner you will have the results, and they will give a more accurate picture of what your problem is. As most of the work in processing the swab is done by a specialist laboratory there is usually a charge to pay. Do not let this put you off as the information which you will get back is priceless when treating your koi. However, avoid routine swab sampling in the absence of any health problems. There is a risk that routine swabbing might result in overuse of antibiotics which will only exacerbate existing problems of antibiotic resistance.

3: Immediately you have finished taking the swab, complete any paperwork which needs to go with it, and get it into the post.

THE RESULTS OF A SWAB

The results of a swab can normally be expected within four to five days. To process a swab the laboratory first smears the end of the sample over an agar plate, which is then placed in a special oven to encourage any bacteria present to grow. Once the bacteria have grown, the strain is identified and then a sensitivity test is carried out. This involves smearing a sample of the bacteria over another agar dish onto which discs are placed which contain different antibiotics. This dish is then put back into an oven and heated for a period before being examined. Any areas which reveal reduced bacterial growth indicate which antibiotics are effective against the bacteria, while any areas where there has been no change or even an increase in bacterial growth show antibiotics that are not effective.

A report is then drawn up by the

laboratory and sent to your local koi specialist as they are normally the account holders for whom the lab works. Depending upon the level of reduction in the growth of bacteria, the relative sensitivity to

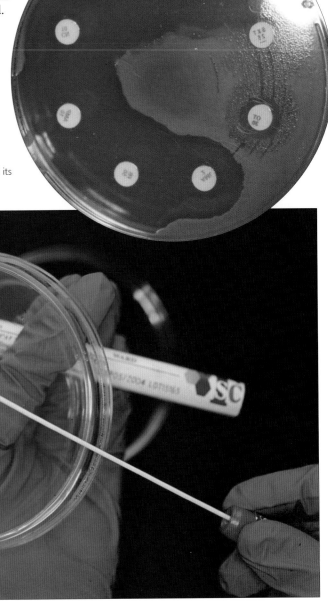

Below and right: An agar plate with eight discs containing different antibiotics (right). A bacterial sample is cultivated on an agar plate (below), and its sensitivity to the different antibiotics observed.

particular antibiotics can be deduced. Although the report may show that a number of antibiotics are effective against the bacteria in question, always try the one which is most effective first. To speed up the whole process this information is generally faxed or emailed, while the original documents are sent in the post. Once the information is received, you will be contacted by your vet or local dealer to discuss the results. From the findings, you will be able to determine which species of bacteria is causing the problem, and at what level it is present. You will also be able to take advice on which antibiotic should be used to treat the problem. In many countries, the use of antibiotics is carefully regulated because of increasing concerns over antibiotic-resistant strains of bacteria and can only be legally prescribed following a veterinary consultation. If the results show a high level of growth of a bacterial species known to cause disease, it may be necessary to administer a suitable course of antibiotic drug. It is now recognized that administering drugs by repeat injection is extremely stressful to koi and this affects their recovery. Most vets will prescribe either a pre-medicated feed or an antibiotic drug that can be mixed with the feed. If smaller koi are infected, it is important to tell both your local dealer and vet that this is the case, as it is not advisable to inject koi that are under 15cm (6in) in length. It may be that an antibiotic food should be used instead.

Remember that although testing water quality and taking a skin scrape should be the first steps taken to identify a disease, a swab should follow if these initial tests turn up nothing. Although many bacterial problems are easy to identify, such as ulceration or mouth rot, it is impossible to tell which species of bacteria is responsible and how it will react to a certain antibiotic. This is why swab analysis is vital as it allows not only the correct diagnosis but also the use of the correct medication. Prolonged use of the wrong antibiotic not only does nothing to help the condition to heal, but it may also lead to the bacteria developing resistance to the antibiotic so making that particular drug less effective in the future. As there are only a limited number of antibiotics available and licensed for use on koi, this is a situation which is to be avoided.

TRANSPORTING KOI

Sometimes you may need to transport your koi to your vet or a koi dealer for further examination to allow an exact identification of a disease to be made, or for extra treatment. While it is always better to treat a sick koi on site rather than putting it through the extra stress of being moved, this is not always possible. So there will inevitably be instances where you have to transport at least one, or maybe more, koi, in order to obtain expert advice or specialist treatment. In order to minimize the stress that this process will cause, it is vital that the koi should be handled and packaged in the correct manner.

Before attempting to transport any koi, make sure that you have the correct equipment needed to bag up your koi with the minimum of stress. You will need suitably sized bags for the koi to be moved, and it is best to have enough of these so that each bag can be doubled up, i.e. one bag placed inside another. This helps to prevent leaks occurring. A supply of air or oxygen will be required to inflate the bag once the koi is placed inside it. Ideally this should be oxygen, but most people do not possess an oxygen bottle. An alternative approach which can be used if the fish will not travel for longer than 60-90 minutes without the bag being opened and resealed, is to use a normal air pump to inflate the bag, and as this is a standard for most koi keepers, it proves a viable alternative. You will also need elastic bands to seal the bags, and a suitably sized box in which a bag can be placed.

Bagging Koi For Transportation

1 Once the koi in question are in the viewing bowl decide on the correct size of bag, and how many koi can be placed in each bag. Plastic koi bags are generally available from your local koi dealer and if you tell them the number and size of the koi that you are intending to move, they will be able to supply the correct size and quantity of bags. Expect to pay for these bags, however, as they are manufactured from thicker plastic, and generally have a double seal to make leaks less likely. These bags come in numerous sizes, but the most common are: 12in x 24in (suitable for small koi up to 5 to 6in), 16in x 30in (suitable for a number of small koi, or individual koi of 10 to 12in), 24in x 40in (suitable for a large number, say up to 30, small koi under 6in, or three to four koi of around 10in

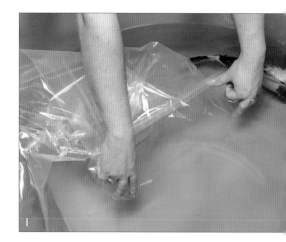

I: Take a fish bag of a suitable size for the koi to be moved.

4: To prevent water from becoming trapped between the two bags, roll the tops of the bags down.

2: Decant a small amount of water into the bag.

3: "Double bag" putting one bag inside another.

5: Coax the koi into the bag adding water while doing so.

6: Having checked that the correct amount of water is in the bag, the top can be unrolled.

7: Hold the top of the bag and one of the bottom corners securely, and then lift it out of the viewing bowl.

to 12in, or individual large koi up to 28 to 30in). For larger koi than this you will require a bigger bag which your koi dealer may not carry as a stock item, and so it may need to be ordered.

2 Select the correct bag size, and add an inch or so of water to it. Then drop this bag inside another bag of the same size. This process is known as double bagging. Before continuing it is a good idea to roll the top of both bags over a

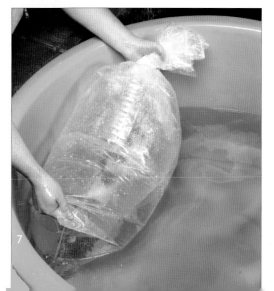

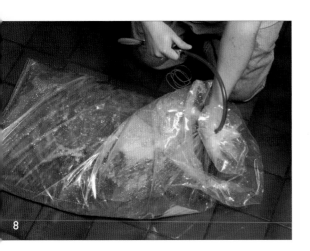

8: Hold the top tight while the air/oxygen hose is inserted.

9: Inflate the bag until only enough plastic is left to tie it off.

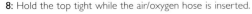

couple of times to prevent the gap between them filling with water. You are now ready to transfer your koi to the bag.

3 Lean over the bowl, take the bag in your hands and gently lower it into the water. As you coax the koi to be moved into the bag, the bag starts to fill with water. Once the koi is in the bag, lift it from the water, and see if it has the correct amount of water in it. For smaller koi the bag should contain about ⅓ water, allowing for ⅔rds air when inflated. However, for larger koi try and ensure that when the bag is laid on its side the water level is high enough to cover the gills.

4 When sufficient water is in the bag, use either an airline from your air pump to inflate the bag if only going a short distance, or if a longer distance is to be travelled use oxygen. Once the bag is inflated, tie it off using elastic bands. It is best to use two elastic bands just in case one should give. It is also vital that not just the inner bag be sealed in this manner, but the outer bag as well.

5 You can place the bagged koi inside a suitable box ready for transporting. Lay the bag horizontally in the box rather than stood up, as this allows the greatest surface area of water to be in contact with the oxygen. This box should be a good fit for the bag to prevent it from rolling about. If the box is too big, pack the space out with something – spare plastic bags inflated with air are a good choice.

6 When placing the box in your car, position the box lengthways across the car. This prevents further damage from occurring to the fish if you have to brake suddenly as the koi will roll in the bag. If the box was positioned the other way, the koi would smash into the ends of the box. This is of extreme importance when larger koi are being transported. It is also an idea to place the box out of direct sunlight to avoid sudden changes in temperature.

7 If you are having to transport your koi over a longer distance, when moving house for example, you might want to consider using polystyrene boxes as these will help maintain a more stable temperature within the box. If it is very hot, you may wish to add ice packs inside the boxes to help maintain a cool temperature throughout the journey.

8 When releasing koi after transportation, check the temperatures of the bag water and the pond water. If they are significantly different, the bag should be allowed to float on the pond for 30 minutes to an hour out of direct sunlight before releasing the koi back into the water.

If you ever need to move a large number of koi it is worth discussing the situation with your local koi dealer, as they may have access to transportation tanks which can be filled with water and put on the back of a truck or trailer. They can also aerate the water in the tank with pure oxygen which is fed from an oxygen bottle on the vehicle via a special

10: Seal both bags separately with elastic bands.

11 and 12: Take a suitably sized box and place the bag in it; then seal the lid securely prior to transportation.

airstone in the tank. Koi moved this way can simply be sock-netted into the tank, then socked out when they arrive at their destination.

Below: Upon arriving at the final destination, you should float the bag on the pond for a period of 30 minutes or so to help stabilize any temperature differences. While doing this, ensure that the bag is kept out of direct sunlight.

KOI DISEASES

This section of the book looks in detail at the various diseases a koi keeper may encounter. Some of the diseases described here, such as tapeworms or *Myxobolus* sp., may occur but rarely cause clinical pathology and are often overlooked, but others like *Trichodina*, whitespot and flukes will undoubtedly be encountered at some time in your koi-keeping hobby. This section is intended to allow readers to identify the problem by using the techniques described in section 2, and then to find the right treatment for that condition. In order to assist you four pages of this section contain an illustrated guide to the diseases discussed. This should be your first place of reference, checking the signs displayed by your koi or on the microscope slide and relating them to the pictures on these pages. From here you can turn to the relevant entry that provides all the information required to make a more educated diagnosis, followed by a recommended course of treatment.

In order to get the best results from this book, and especially this section, it is vital that you should read section 2 closely and follow all the suggested steps to ensure that a correct identification of a disease is made, either by visual inspection, taking a skin scrape, or sending away a swab. If you do not take care doing this, an incorrect diagnosis may be made and the wrong treatment applied. This can be just as harmful – or even more harmful – than no treatment at all. That is why it is vital for a koi keeper to own a microscope, and to be prepared to send swabs away for processing when bacterial problems are suspected. Your local koi dealer or specialist will be able to help with this, although do not expect to get these services for free as they do cost money. However, this outlay is very small when compared to the cost of replacing an averagely stocked koi pond!

As well as making the correct identification, it is vital that you maintain a medicine chest containing appropriate treatments. Although it is impossible to keep stock of every treatment for every condition all the time, a few simple items will help with a vast number of the common conditions and complaints experienced by the typical koi keeper. Try and have the following to hand as a "just in case koi first aid kit" – sedating agents, malachite green, formalin, potassium permanganate, salt (this must be cooking salt, not table salt), Acriflavine, propolis or a similar wound sealer, cotton-wool buds or similar, a sharp pair of scissors, a scalpel and, of course, a good net and bowl in which to catch and handle the koi. Your local koi dealer should be able to supply these items, and may even sell an off-the-shelf first-aid kit containing most, or all, of them. These few items alone will stand you in good stead to combat a wide range of diseases, and respond quickly once an exact diagnosis has been made.

As you consult this section, it will soon become apparent that a few simple measures can help prevent the vast majority of the diseases featured. Good water quality, regular system maintenance, the use of a good quality food, pond heating, and a sensible low stocking level (see page 11) are the most effective measures to prevent disease from occurring in the first place. If you bear these precautions in mind and spend a bit of time each week ensuring that your system is maintained in optimum condition, there will be far fewer occasions when you will need to call upon this part of the book. However, when those times do arise, all the necessary information to progress from exact diagnosis to sensible treatment can easily be found and understood. This guidance allows you to pursue an educated treatment programme so as to return your pond to full health in the shortest possible time.

Below: The koi in this picture has eye damage and damage to the gill cover which may be susceptible to secondary infection, and so may require topical treatment.

The eye and gill damage on this fish could have been caused by physical damage or by a parasite infection, such as *Trichodina*.

PARASITES, VIRUSES, BACTERIA AND FUNGI

Parasites are organisms that require another organism, known as the host (i.e. your koi), in order to complete their life cycle. Parasites are detrimental as they either feed off the host, or alter the behaviour to favour the parasite so it can infect it's preferred host. External parasite infections in koi are commonplace and can

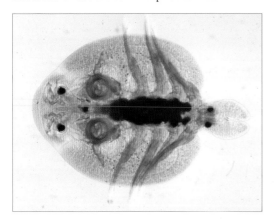

Above: A close-up of *Argulus* (a fish louse). This is one of the few parasites which can be seen with the naked eye.

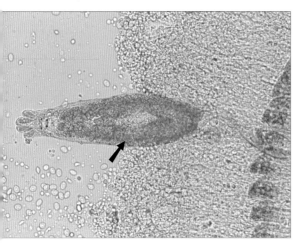

Above: *Dactylogyrus* is a fluke that infects the gills. It is here seen under magnification through a microscope.

generally be identified by behavioural or visible changes to the infected koi, or by taking a skin scrape and making an exact identification under a microscope. Diseases caused by internal parasites are much harder to identify because they live inside the fish. Generally most external parasitic infections can be treated easily once an accurate diagnosis is made. It is normal to find more than

one species of parasite at low levels amongst any population of koi. It is quite normal to find one or two types of parasites present at any time in most ponds. It is only when your koi become stressed that their natural resistance drops and you will see the levels of parasites rise and start to cause a problem.

What Is A Virus?

Viruses are unique organisms, comprising a piece of genetic material surrounded by a protein coat. Whether they are truly 'living' is controversial, as they do not feed and are completely dependent on a host in order to reproduce. Upon infecting a cell, the virus takes over the mechanisms and diverts the cell from making proteins, hormones, lipids, etc. to producing more viruses. Infection by viruses leads to cell death and organ damage, affecting the function, which may ultimately cause the death of the koi. The immune system mounts a response to the infection; usually the virus is eliminated and the koi recovers. When certain viruses infect a cell, they insert a copy of their genetic material into the host cell's genes. Examples include herpesviruses, which cause carp pox and Koi Herpesvirus (KHV). It means that infected koi become carriers of the virus. The carrier fish are fit and healthy, but if subjected to stress, the virus can be shed and an outbreak of disease occurs.

There is no treatment for a virus, but outbreaks of disease can be prevented by vaccination. Vaccines are prepared using either dead whole virus, specific sections of the protein coat or attenuated virus, which although live is engineered to make it either less harmful or unable to cause disease. The specific immune system responds to the challenge created by the vaccine and antibodies are produced that recognize and suppress the disease-causing virus. Following vaccination, the fish immune system remains responsive to viral challenge for about two years.

What Are Bacteria?

Bacteria occur in a number of shapes and sizes, but all are microscopic, single-celled organisms, with a simple strand of genetic material that lies free within the cell. In higher organisms, the genetic material is contained within a membrane inside the cell. In addition to the strand-like genome, there are also small additional pieces

Above: A bacterial infection is shown here. This may have originally started as physical damage caused by flicking.

ubiquitous species is found in every aquatic environment, but is an opportunist pathogen, taking advantage of weak or stressed koi, particularly if there is damage to the epidermis. Once this outer layer of skin has been abraded or damaged in a weak or stressed koi, bacteria rapidly colonize the wound, which can lead to unsightly ulceration. Outbreaks of ulceration within a population of koi generally indicate that the koi have become stressed, allowing bacteria to infect any slight wound. Ideally, you should ask a vet or fish health specialist to identify the bacteria causing the disease and to give the koi appropriate medication.

of genetic material known as plasmids, which reproduce separately from the main chromosome and are readily transferred between bacteria, even unrelated species. The genetic material within the plasmids enables bacteria to develop resistance to drugs such as antibiotics. Bacteria can reproduce by simple division, which means the numbers in a colony can increase very rapidly and reach millions in a few hours, but they can also reproduce sexually. Transferring genetic material in the plasmids enables bacterial resistance to antibiotics to spread quickly.

Bacteria are naturally present in a pond. Indeed, the biological filtration system is reliant on the beneficial nitrifying species to oxidize ammonia to nitrate. In addition to the nitrifying species, many hundreds of free-living species of bacteria are found in the koi pond and live in harmony with your koi. The koi intestine harbours its own bacterial flora and these not only help the koi to digest food, but even assist in destroying species with the potential to cause disease. In most instances, stressed koi are more susceptible to bacterial infections, but true disease caused by bacteria can only be caused through introduction of an infected or carrier koi.

Koi are more likely to become infected with bacteria such as *Aeromonas hydrophila*. This

What Is Fungus?

Although there has been an assumption that fungi are plants, this group of organisms is now considered a discrete taxonomic group, more closely related to animals than plants. Fungi live in wet or damp conditions and are either parasitic, taking nutrients from a living host to complete their life cycle, or saprophytic, feeding on dead and decaying organic material. Fungus tends to grow by means of branching threads, rather like the roots of a plant, and reproduction takes place both sexually and asexually. Both result in the release of spores, which are the stages that can infect koi. The most commonly encountered fungus in the koi pond is familiar as cottonwool-like tufts. Although white, they are usually a dirty brown colour, as the branching threads trap pieces of debris and dirt. A number of fungus species cause disease. Generally, these outbreaks of fungus are associated with physical injuries to koi or fish that are in a weak and debilitated condition. Whilst most koi keepers are concerned with fungi that cause disease, there are many free-living species that play an important part in assisting in the breakdown of organic material in the biological filtration system.

DIAGNOSTIC GUIDE FOR CONDITIONS CAUSED BY PARASITES

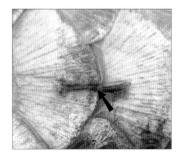

Anchor worm: *page 66*

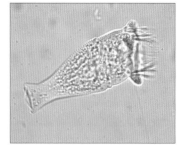

Apiosoma: *page 70*

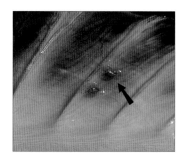

Argulus (fish lice): *page 72*

The gills and area around the eyes are susceptible to parasite infestation.

Parasite infections can lead to heavy mucus production.

Chilodonella: *page 76*

Some parasites, like tapeworms, live inside the fish and so are not visible externally.

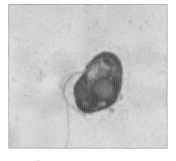

Costia: *page 82*

Epistylis: *page 92*

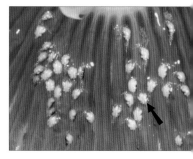

Gill maggots: *page 98*

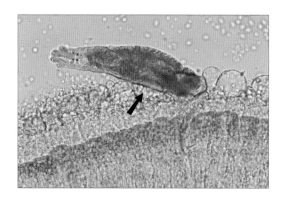

Gill and skin flukes: *page 102*

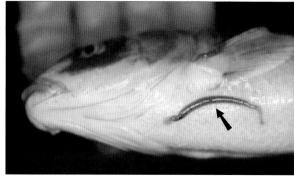

Leeches: *page 110*

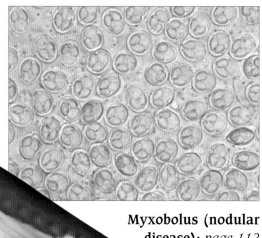

Myxobolus (nodular disease): *page 112*

Tapeworm: *page 120*

Whitespot may even be found on the tail.

Parasites like leeches and anchor worm may be seen attached to a fish's body.

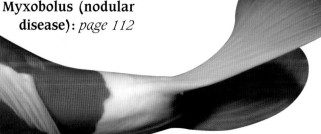

Trichodina: *page 122*

Whitespot: *page 124*

DIAGNOSTIC GUIDE FOR CONDITIONS CAUSED BY BACTERIA, VIRUSES, FUNGI AND OTHER EXTERNAL AND INTERNAL DISORDERS

Bubbles in the fins are an indication of gas bubble disease. Fins are also vulnerable to fungal rot.

Ulcer disease: *page 62*

Columnaris: *page 78*

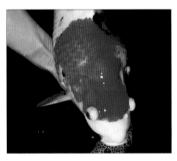

Curvature of the spine: *page 84*

Dropsy: *page 86*

Egg retention/Internal tumours: *page 88*

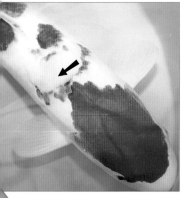

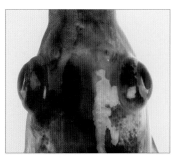

Gas bubble disease: page 96

Gill disorders: page 98

Fungal disease: page 94

Skin lesions are often the result of flicking to relieve the irritation caused by other infections.

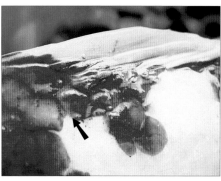

Hi-kui: page 106

Conditions like dropsy cause external lifting of the fish's scales so that it resembles a pine cone.

KHV and SVC: page 108

Papilloma and carp pox: page 114

Pop-eye: page 116

Swimbladder disorders: page 118

ULCER DISEASE

An outbreak of ulcer disease in a koi pond may be due to infection with a number of different species of bacteria or even coronavirus, which has been implicated. Most commonly, the primary agent in an outbreak of ulcer disease is atypical *Aeromonas salmonicida*, a species of bacteria that is a true fish pathogen, infecting the skin and initiating an ulcer. Once the skin has been breached by the atypical *A. salmonicida*, the lesion is rapidly colonized by *Aeromonas hydrophila*, a related but ubiquitous species found in every aquatic environment. In most instances, laboratory results from swabs taken from an ulcer or lesion reveal only *A. hydrophila*, as the atypical *A. salmonicida* is difficult to isolate. *Flavobacterium columnare* may also prove to be a problem for the koi keeper, leading to fin rot and ulcers. Whichever species of bacterium is suspected, it is important to try to have the infection identified by a laboratory in order to ascertain an appropriate course of treatment.

Identification

Signs of infection with bacterial disease are variable. Often, there may be an initial display of irritability, with twitching fins or repeated rubbing and flicking, but samples of mucus fail to show any parasite infection. The more obvious signs are lethargy, when the koi is very inactive, accompanied by loss of appetite. If the koi is stressed or in any way debilitated, once its skin has been damaged, either by physical damage, other pathogenic bacteria or even viral infection, it will quickly be infected. Early signs of infection are localized areas of inflammation, reddening and sometimes areas of raised scales. All too often, koi keepers only become aware of bacterial infections when there is an ulcer on the body.

In most instances, the recommended course of action is to take a swab from the infected koi, but it is important to bear in mind that laboratory results are likely to identify *A. hydrophila*, even though this is an opportunist pathogen and not necessarily the causative agent. Nonetheless, even if the underlying infection is not identified, the presence of *A. hydrophila* may need treating in order to contain the ongoing infection. Most

Right: This koi has a severe case of *Aeromonas*. It has resulted in lifting of the scales, which in turn may be an indication of bacterial dropsy.

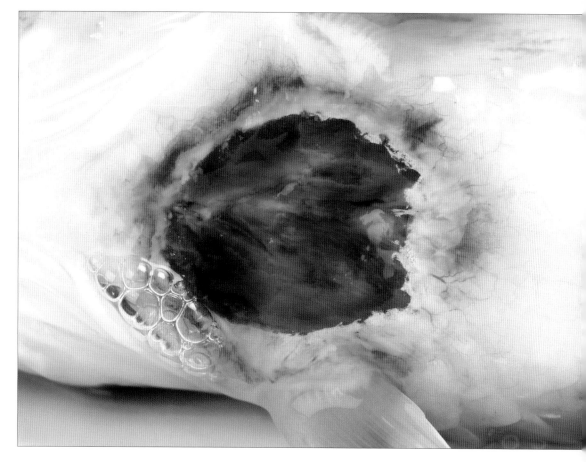

Above: An ulcer on the underside of a koi. This type of bacterial infection can be hard to spot until it has reached an advanced stage, due to its physical location.

laboratory services will include a sensitivity test, where colonies of the bacteria isolated from the infected koi are plated with antibiotic drugs to test the degree of resistance.

Prevention

The best way to reduce outbreaks of opportunist bacterial disease is to run a clean pond. It is impossible to eliminate bacteria completely from the water, so it is vital to ensure the environment in which your koi live is kept to the highest standard to reduce stress to a minimum. Bacteria such as *Aeromonas hydrophila* are found in every aquatic environment, but thrive in koi ponds where filters are poorly managed and drains are flushed infrequently, leading to large quantities of particulate organic material. Koi ponds with low stocking levels are readily managed and the low numbers of koi lead a stress-free life as there is little competition for essential resources such as food and oxygen.

There are various means of reducing the bacterial load on a koi pond, such as using ozonisers or even ultraviolet sterilizers, but a well-run koi pond should not be subject to large numbers of bacteria. Ultraviolet sterilizers are designed for fish hatcheries, where it is important to keep the bacterial load to a minimum because larvae and fry have little immune protection and disease spreads quickly amongst large numbers of fish. These ultraviolet sterilizers differ from the UV clarifier systems commonly used to control green water caused by suspensions of algal cells. UV sterilizers comprise multiple units through which the water flows and the water is exposed to the light for a much longer period. Realistically, there is no need for such powerful sterilizing units on a well-managed koi pond.

Treatment

For any Aeromonas or other bacterial infections, treatment takes the form of anti-bacterial drugs which normally have to be obtained from a vet. However, for minor infections alternative approaches, such as herbal (Melaleuca) extracts to deal with small bacterial ulcers, may be considered along with the possible use of off-the-shelf proprietary bactericides, which will be available from your local koi dealer.

Sometimes, despite using off-the-shelf medications, there really is little improvement in the condition of koi infected with an ulcer, or other koi in the pond are also affected. Persistent ulcers or an increase in the numbers of koi affected may need a stronger medication than

Below: This koi with a bacterial infection on the caudal peduncle has lost a number of scales. The white patch is an area of necrotic tissue and the surrounding area is inflamed.

any of the proprietary treatments.

This is the time to involve your local vet, as it is likely the koi will need a course of antibiotic treatment. The vet will probably arrange for swabs or other samples to be sent for microbiology and sensitivity testing to an appropriate laboratory. The results of the sensitivity testing will enable the vet to prescribe the most suitable course of antibiotic medication.

Antibiotic drugs are powerful medications; indeed, some may have certain quite harmful side effects that may exceed any beneficial activity, so they should be administered for the appropriate koi body weight and for a specified period of time. It is usual for antibiotic drugs to be veterinary prescription medicines and as such there is often a legal requirement for a vet to examine the koi in order to dispense the most suitable medicines.

These days, there is a preference for feeding antibiotic treatments to koi as the least stressful

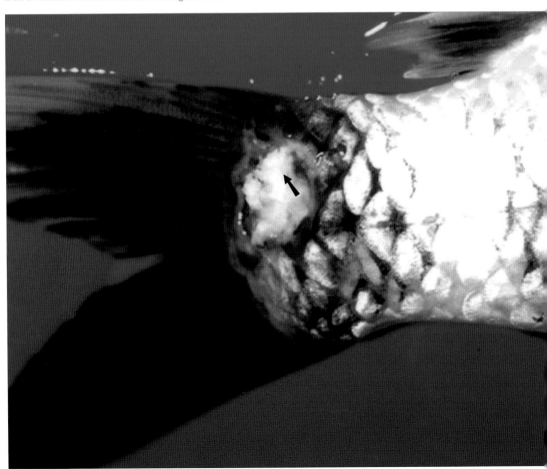

way of administering a course of treatment. Occasionally a koi may be too sick to feed, in which case the vet may need to sedate it and administer the antibiotics by injection. It is important that administration of antibiotics is supervised by a vet, who will understand the correct dose rates and the biochemistry of the drug in the animal under treatment. Incorrect doses of antibiotics and failure to complete a full course of treatment are the means by which bacteria become resistant to these drugs, which are essential to human and animal medicine.

Regrettably, many of the bacterial diseases that infect koi are now multi-drug resistant. It may be difficult to find a suitable single treatment and it may be necessary to use a combination of drugs in addition to the new range of antimicrobial disinfectants. These may be used in the pond environment and are very effective at controlling many infections, without the necessity of using drugs such as antibiotics. Disinfectants such as Virkon S Aqua are extremely effective at reducing the bacterial load in a koi pond. Traditional treatments, such as acriflavine and proflavine hemisulphate have always been used to assist healing during outbreaks of ulcer disease. Other treatments such as potassium permanganate may be used at the dose of 1.5g per 1,000 litres (220 gallons). Use this chemical with caution in hard water areas as there is a tendency for manganese oxide to coat the gills, which disrupts breathing. Chloramine T is also a useful antibacterial at the minimum dose of 1g per 1,000 litres (220 gallons), but use it with extreme caution in soft water, where it is toxic.

Below: When bacterial infections occur, scales may die. Unfortunately, these may have to be removed to allow the infection to heal and prevent it from spreading.

ANCHOR WORM (LERNAEA)

This infection is caused by the parasite *Lernaea cyprinacea*, which is actually a crustacean that superficially resembles a worm. The head forms an x-shaped anchor, hence the common name, anchor worm. Mature anchor worms are visible with the naked eye as numerous thick, hairlike strands protruding from the koi's body and are 5–12mm (0.2–0.5in) long. The female is parasitic, laying eggs in two sacs that are shed into the water. On hatching, the anchor worms are typical copepodid crustaceans and undergo a free-living stage. Prior to sexual maturity, the copepodid anchor worms seek a koi host and make their way to the gills, where they attain sexual maturity and mate. After mating, the males die and the females move onto the koi's body, where they burrow beneath a scale or even directly into the skin of doitsu fish. The female's head transposes into the characteristic anchor, the legs become vestigial as the body elongates and the female feeds on the koi blood and body fluids, gaining nutrients to produce more eggs.

Identification

Lernaea will infect a koi in many places but the most common sites for infection are the body (practically under the scales), the mouth, around the eyes, on the fins, the gills, and the joints of fins where they meet the body. In the very

Right: Dead anchor worms which have been removed from a fish. The larger end is what actually attaches to the koi.
Below: An anchor worm's attachment organ fixes itself by physically penetrating under the fish's scale.

The life cycle of the anchor worm
(Lernaea)

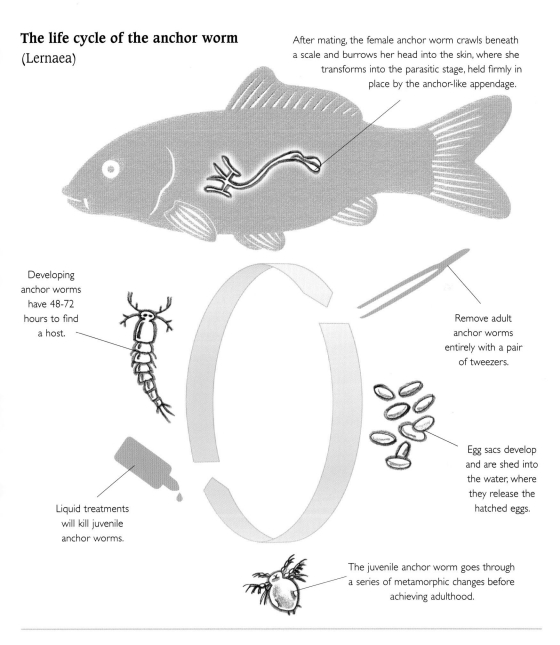

After mating, the female anchor worm crawls beneath a scale and burrows her head into the skin, where she transforms into the parasitic stage, held firmly in place by the anchor-like appendage.

Developing anchor worms have 48-72 hours to find a host.

Remove adult anchor worms entirely with a pair of tweezers.

Egg sacs develop and are shed into the water, where they release the hatched eggs.

Liquid treatments will kill juvenile anchor worms.

The juvenile anchor worm goes through a series of metamorphic changes before achieving adulthood.

early stages of an anchor worm infection it is not possible to spot the typical anchor worm shape, and so infection may go unnoticed or be wrongly diagnosed, as it can look as if the koi has another parasite infection, such as whitespot, because of the presence of white dots on the body. These are in fact the young anchor worms. This is why it is vital to carry out a skin scrape when any disease is suspected to ensure that a correct identification is made and the right course of treatment applied.

A koi infected with anchor worm may appear to move erratically and even rub or flick against any hard surfaces in the pond to try to relieve the skin irritation and dislodge any anchor worms attached to its body.

Anchor worm infections can reach quite high levels before they cause any real threat to the koi, except when the mouth or gills are being infected, as this may affect the koi's ability to breathe or eat. Another problem is that the attachment site normally becomes infected either by bacteria, fungus or even a virus, because the anchor worm punctures the skin to feed on tissue fluid and cells, leaving an area prone to infection. Characteristic signs are scales lifting and the area around the anchor worm becoming red and inflamed. As the anchor worm penetrates beneath the koi's skin, it is common for secondary infections – usually bacterial – to occur. The bacteria readily infect the tissues damaged by the anchor worm, leading to ulceration. The life cycle of the anchor worm is affected by temperature; at low temperatures egg development and maturation of the copepodid larvae slow down.

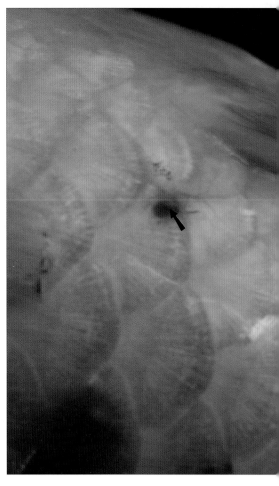

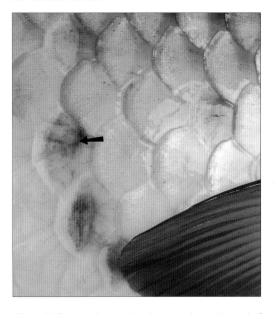

Above: Unfortunately removing the worm is not the end of the story. The attachment site often requires treatment. This is a typical example of anchor worm damage.

Prevention

Anchor worm is a common problem on new imports of koi, especially when they have been recently harvested out of mud ponds in which they have spent the summer growing. It is quite common to find some anchor worm on new imports of koi, and most dealers will check for their presence and treat as a matter of course if any are found. Therefore, it is advisable to wait for at least a couple of weeks before taking delivery of koi which are newly imported to allow any treatment to take effect. However, in recent years Japanese koi breeders have addressed the problem of anchor worm and it is less common for koi to be infected.

Treatment

If you find anchor worm, first remove the adults physically from the koi. Sedate the fish, then gently pull the anchor worm from the koi with a pair of sterilized tweezers. Apply a suitable topical treatment, such as malachite green and propolis or just propolis, to the attachment site. To get the full benefit from doing this, be sure to inspect all the koi in the pond at the same time and repeat the procedure after ten days or so to ensure that no new anchor worm have attached themselves or that others have been missed. At the same time, check if any secondary bacterial infection has occurred and treat it accordingly, seeking advice from your vet or local koi health specialist.

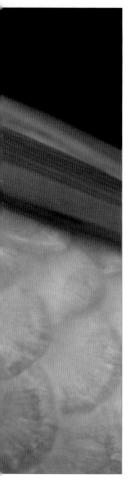

Above left: Although appearing very faint, a small anchor worm can be seen here attached to the underside of a koi close to its pelvic fins.

Above right: Here the anchor worm is being removed. If tweezers are not available, you can push the anchor worm out with your fingernail, but be sure that it comes out whole and that no part is left still attached to your koi.

It is also vital to treat the pond to eradicate the free-living stages of the anchor worm parasite. This will need to be done several times, as the treatments available do not always penetrate the egg sacs. The development of the eggs is slower at lower temperatures – for example, at 16°C (61°F) they take approximately six weeks to hatch. It may be sensible to treat the pond at six-weekly intervals for several treatments to ensure all developing stages are effectively eradicated. It may prove difficult to find a brand of medication that will treat anchor worm effectively.

The most effective treatment for anchor worm is based on emamectin benzoate and there are formulations coming onto the market that are suitable for treating infected koi. In most instances these products are only available from a vet. It is certainly worth seeking professional veterinary advice with regard to the use of this drug as it is potent and the dose needs to be administered very carefully. Certain fish species, such as rudd *(Scardinius erythrophthalmus)* and orfe *(Leuciscus idus),* are very sensitive to the product and should be removed from a pond prior to treatment. Examine these fish carefully for anchor worm infection and remove any adult female anchor worms. Failure to remove any adult anchor worm from sensitive species would result in the pond becoming reinfected when the rudd or orfe are returned to the koi pond.

APIOSOMA

Although often described as a parasite, *Apiosoma* is actually a commensal, which means it derives a benefit in terms of food by attaching to the koi, but the latter neither gains nor suffers from the relationship. *Apiosoma* is flask shaped, with a ring of tiny hairs, known as cilia, around the mouth. The cilia waft food in the form of particulate organic material, such as dead cells, sloughed mucus and bacteria, into the mouth of the *Apiosoma* and it is not uncommon to find them in large numbers near open wounds or ulcers on koi. In order to remain fixed onto the koi, *Apiosoma* secrete a tiny pad of a glue-like substance, which may be an irritant to the koi. *Apiosoma* pose the greatest hazard to fry or juvenile koi, as they can coat the gills, affecting breathing and osmoregulation.

Identification

External signs of an *Apiosoma* infestation include the production of excess mucus which results in the skin looking opaque. In severe cases the actual skin colour may disappear as the mucus becomes so thick the body actually looks white. Infected fish may start to hang in the water with their fins clamped, and spend more time in areas of heavily oxygenated water, such as around water returns and airstones. This will especially be the case if the gills are infected. If an extremely large quantity of *Apiosoma* are present, koi may flick their heads from side to side to

Above: One of the characteristic physical symptoms of *Apiosoma* is emaciation.

Above centre: *Apiosoma* is not visible to the naked eye and so a microscope is required to make an exact identification of this protozoan.

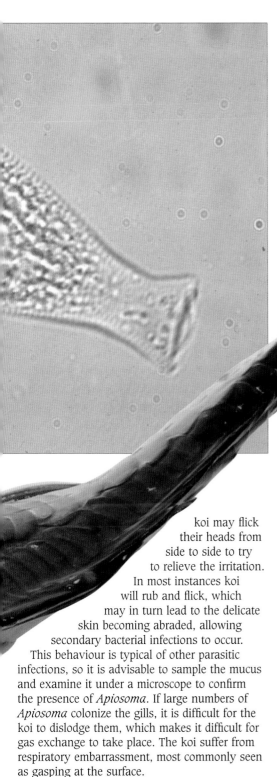

Prevention

Given that *Apiosoma* is a commensal that feeds on particulate organic matter and bacteria, it is unlikely to be encountered where the pond and filtration system are well maintained. It is only when environmental conditions are allowed to deteriorate that a problem arises. To prevent *Apiosoma* infecting koi, it is important that the water is kept free of organic material by regularly flushing drains and removing the remains of any uneaten food. Maintaining the filtration system is one of the most important tasks related to reducing organic material. The mechanical filtration, such as brushes or the first vortex chamber in the filter, are designed to remove the organic material from the water and therefore should be kept clean and free of detritus. In the summer months when the koi are more active it is very important to raise standards of hygiene and maintenance in order to keep the pond and filters clean.

Treatment

Less severe cases of *Apiosoma* can be easily treated with a proprietary anti-parasite medication or the use of malachite green and formalin – the dose rate will depend upon the concentration of the mix. Alternatively, you might consider using potassium permanganate at a dose rate of 1.5g per 1000 litres (220 gallons), which can be repeated every five to seven days over a period of three weeks to a maximum of three treatments. Use with caution in hard water areas.

Salt can also be an effective treatment, used either as a bath or a pond treatment. If the infection has become severe, the chances are that there will be numerous sites of secondary infection on the fish and these will need to be topically treated with propolis. Although treatment will normally result in complete eradication of the parasites, a severe gill infestation may cause gill damage and so further losses of fish may take place even after the problem with the parasite has been eliminated.

koi may flick their heads from side to side to try to relieve the irritation. In most instances koi will rub and flick, which may in turn lead to the delicate skin becoming abraded, allowing secondary bacterial infections to occur. This behaviour is typical of other parasitic infections, so it is advisable to sample the mucus and examine it under a microscope to confirm the presence of *Apiosoma*. If large numbers of *Apiosoma* colonize the gills, it is difficult for the koi to dislodge them, which makes it difficult for gas exchange to take place. The koi suffer from respiratory embarrassment, most commonly seen as gasping at the surface.

ARGULUS (FISH LICE)

Despite their name, fish lice are crustaceans and related to the more familiar crabs and lobsters. Unlike other parasites that spend their lives associated with the koi, fish lice are very adept swimmers, even though they spend long periods attached and feeding on the koi. Fish lice feed by means of a special organ known as the stylet, which pierces the skin and injects an anticoagulant, allowing the parasite to suck blood and fluid from the koi. The anticoagulant is an irritant and also contains a chemical that attracts other fish lice to feed around the same site. This is why large numbers of them are often found together. The injuries caused as the fish lice feed often become secondarily infected with bacteria, which can also lead to ulceration. Because of the way fish lice feed by puncturing the skin and sucking blood and body fluids, bacterial infections are rapidly spread between a population of koi. It is also possible for fish lice to transmit viral disease among koi. Fish lice tend to congregate where the koi find it difficult to dislodge them, such as around the bases of the pectoral and pelvic fins and mouth. The female fish louse swims off the koi and lays her eggs amongst plants, plant baskets, aquatic soil, blanket weed or even in the filtration system. The development of the eggs is temperature dependent; at about 16°C (61°F) it takes approximately six weeks for the eggs to hatch, but at higher temperatures development is much more rapid. Conversely, at lower temperatures it is significantly longer. As

Below: Two *Argulus* attached to the dorsal fin of a koi.

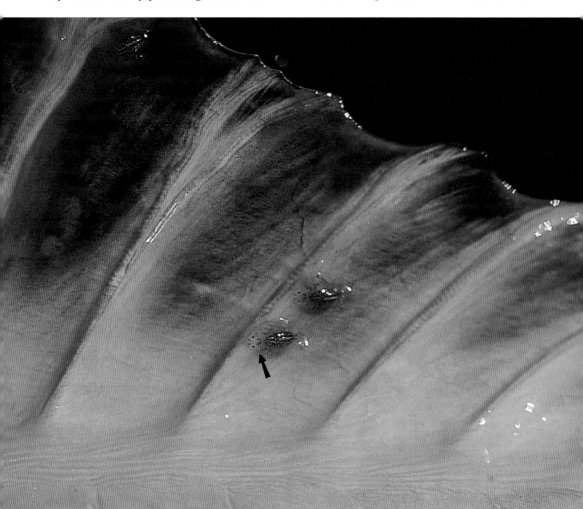

The life cycle of the fish louse (Argulus)

Fish lice moult numerous times and mature individuals cover a range of sizes.

Mating can take place both on the koi and in the pond environment. A single mating is enough to fertilize all the eggs ever laid.

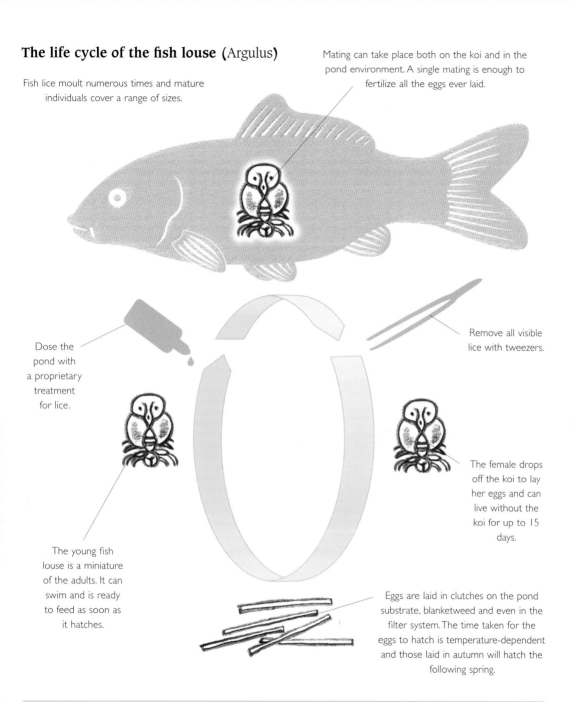

Dose the pond with a proprietary treatment for lice.

Remove all visible lice with tweezers.

The female drops off the koi to lay her eggs and can live without the koi for up to 15 days.

The young fish louse is a miniature of the adults. It can swim and is ready to feed as soon as it hatches.

Eggs are laid in clutches on the pond substrate, blanketweed and even in the filter system. The time taken for the eggs to hatch is temperature-dependent and those laid in autumn will hatch the following spring.

soon as the young fish lice hatch, they look like miniature adults and will seek a koi to commence feeding. As the fish lice grow, they undergo a series of moults but continue to feed each time on the koi. Once they attain sexual maturity, the fish lice mate and the life cycle is repeated.

Identification
Fish lice can be easily seen with the naked eye, and appear as semi-transparent discs swimming in the water or attached to a koi. They range in size generally from 3-9mm (0.1-0.35in), although in some instances may be as large as

12mm (0.5in). A koi infected with lice will show a number of symptoms. Even with moderate infections it is possible to see the lice attached to the koi. A koi suffering from infection with fish lice will classically display signs of irritability: the fins may repeatedly twitch and the fish will

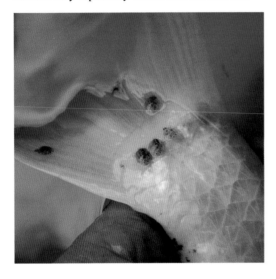

Above: *Argulus* can be seen without a microscope, either whizzing around in the water or on the fish themselves.

flick and rub against pond surfaces. In extreme cases, koi will swim with two-thirds of the body out of the water, using the tail to propel them along, a motion described as similar to a dolphin performing a trick. The rubbing and flicking can potentially abrade the skin, but the greatest injury to infected koi comes from the lice, which damage the skin. The initial response of the skin is to thicken, rather like a callus, in response to the repeated bites. The feeding mechanism of the lice not only punctures and damages the skin, but also introduces bacteria into the wound and this often results in the formation of unsightly ulcers.

Prevention

The only way that fish lice can be introduced to a koi pond is by stocking an infected fish. When you buy a new koi, most reputable retailers will allow you to have a close look at the fish. Ask to see the koi in the plastic transport bag and look at it not just from the top but hold up the bag so that you can look underneath to see the belly and fin bases. If fish lice are present it is usually possible to see them with this brief inspection.

Treatment

In an open lake, fish lice live around the margins, in the littoral zone. Although they are capable of swimming, they do not venture into the deeper water so must wait for the fish to come to them. The size of the water and the infrequency of finding a host keeps the fish louse populations in check in the wild. In the koi pond, the smaller volume of water, coupled with the availability of a fish host, strongly tips the balance in favour of the fish louse. If you spot a koi in your pond with fish lice, the chances are that most, if not all, the koi in your pond will be infected. It is important that all the koi are checked for the presence of this parasite. Many koi keepers want to remove the fish lice physically, but it is usually necessary to tranquilize or sedate the koi before removing any of the lice using a pair of clean tweezers. Once the fish lice have been removed, treat the affected area with a topical antiseptic appropriate for use with fish. Physically removing the fish lice from the koi allows the koi keeper to ensure that there are no secondary infections arising from the feeding parasites. However, it is very stressful for the koi. Assuming the fish are feeding, they can be viewed while eating to watch for any signs of ulceration.

The problem with fish lice is that as well as the lice present on the koi, there are also large numbers in the pond water, on surfaces and filters, all of which need to feed on the koi. It is therefore essential to control the fish louse population in the whole pond. Various treatments are available over the counter which indicate they will control fish lice infections in the pond. More recently, treatments based on emamectin benzoate have been formulated into water soluble treatments and are available on veterinary authorisation. These new treatments will eradicate both the fish lice in the pond and those on the koi, which is beneficial as it eliminates the need to anaesthetize the koi and remove the lice by hand. As the life cycle of fish lice is temperature dependent, it is probably sensible to stagger a treatment of three separate doses at intervals of about six weeks.

Right: *Argulus* viewed under a microscope. As a koi keeper you would not see this detail when inspecting your pond, just small translucent discs with black dots inside them attached to your fish.

CHILODONELLA

Chilodonella is a single-celled, heart-shaped, microscopic organism that infects the skin and gills. It moves by means of tiny hairs (cilia) arranged in rows along the longitudinal axis. These beat rhythmically and propel the *Chilodonella* along. *Chilodonella* feeds using a group of stiffened cilia called the oral basket. It inserts this into the koi's skin cells, sucks out the cellular contents and consumes them. *Chilodonella* prefers cool water of about 10°C (50°F). Heavy infections of *Chilodonella* cause koi to produce excess mucus, but the skin cells are both damaged by the parasite and become exhausted from copious mucus production. Often, the body surface of heavily infected koi becomes dry and secondary fungal or bacterial infections readily occur. Infection with *Chilodonella* is through contact and it spreads rapidly in an overstocked koi pond.

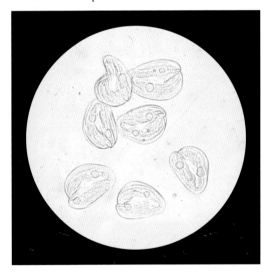

Above: For an exact identification of the protozoan *Chilodonella* to be made, a microscope must be used.

Identification

Koi with a moderate infection of *Chilodonella* tend to show signs of irritability, such as frequent rubbing or flicking on the surfaces of the pond or even jumping out of the water. This behaviour can lead to abrasions or injuries, which may then become secondarily infected with bacteria or fungus. Often, the irritation caused by the *Chilodonella* leads to the production of excess mucus on the body, which appears rather like a grey film, either patchy or covering the whole body. The koi may also appear 'stressed', a term commonly used by koi keepers to describe the appearance of prominent blood vessels clearly visible on the skin of the fish. Even moderate infections of *Chilodonella* can result in quite significant damage to the skin and gills because of the invasive way it feeds. Left untreated, outbreaks of *Chilodonella* can result in secondary infections as a consequence of the parasite damage, or to the condition known as dropsy. The integrity of the skin and gills is affected and this in turn compromises osmoregulation.

When koi display signs of irritability it is always important to determine whether the cause is environmental (for example, high nitrite concentration will also cause the fish to rub on the pond surfaces and jump) or if it is a result of a parasite infection. It is also important to identify correctly the parasite that is causing the irritability and to treat the koi with the appropriate medication. Parasite identification is achieved by taking a sample of mucus from several of the koi and examining these under a microscope. If a single koi has a moderate to severe *Chilodonella* infection, it is feasible that there is an underlying, undiagnosed problem that is causing it to become debilitated and the parasites are simply taking advantage of a weak fish. Where many or all the koi in a pond are affected, there is usually some other environmental issue that is causing the fish to become stressed and therefore susceptible to infection with *Chilodonella*. As this parasite is prevalent at low temperatures, it could simply be that the koi have become stressed by a prolonged period of cool water conditions.

Prevention

It is fair to say that within a koi pond, parasites will always be present at very low levels on most of the koi. Fit and healthy koi are able to regulate the number of parasites through their immune system, but when they are stressed, debilitated or unwell, then the parasites reproduce unchecked, leading to moderate to severe infections. If the pond is heavily stocked rapid infection will also occur, so sensible stocking levels should be observed. Good water quality and system husbandry will also help in preventing an outbreak of this disease as it is only at times of stress that the parasite will be able to multiply at a rate quick enough to become a major concern.

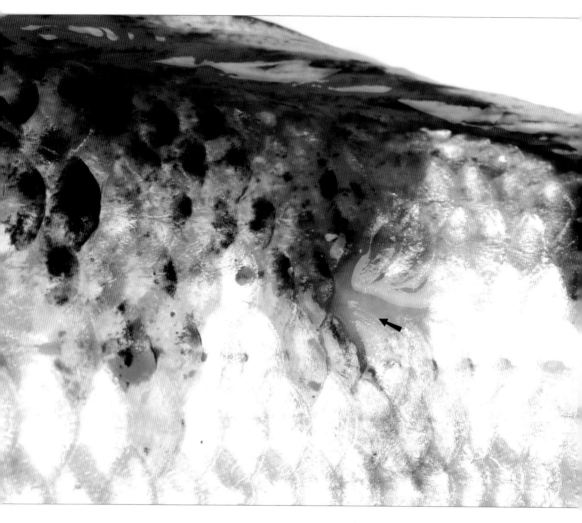

Above: *Chilodonella* causes infected fish to flick against the side of a pond, which may lead to physical damage like this.

To help reduce stress the use of a pond heating system is also desirable to prevent fluctuations in temperature as this disease is more common in the spring and autumn when the water temperature warms after winter and then cools as winter approaches. *Chilodonella* is a greater threat at lower water temperatures of about 10°C (50°F). If the water temperature is kept above this, and good husbandry techniques are applied, the chances of a major outbreak are reduced.

Treatment

Chilodonella can be treated with one of the proprietary brands of manufactured anti-parasite treatments. Alternatively, some koi keepers prefer to prepare a mixture of malachite green and formalin (Leteux-Meyer mix). Combining malachite green with formalin has a synergistic effect; in other words, it creates a more potent and effective treatment when the two are mixed. Take care when making the Leteux-Meyer mix, as the volume of malachite green used is considerably less than if this chemical is used alone. Potassium permanganate may be considered at a dose rate of 1.5g per 1000 litres (220 gallons), but use it with caution in hard water areas. Salt may also be used to control *Chilodonella*, but not in conjunction with either formalin or potassium permanganate as by-products are formed when these are used together that are very harmful to the koi.

COLUMNARIS DISEASE (FIN ROT, GILL ROT, MOUTH ROT/COTTON WOOL AND SKIN COLUMNARIS)

*F*lavobacterium columnare (previously known as *Flexibacter columnaris*) is associated with infections of the skin and gills of koi. This species of bacteria is found in every aquatic environment, as well as on the koi, and prefers temperatures higher than 15°C (59°F). The warmer the water, the more quickly this bacterium infects susceptible koi. High organic loads – and particularly uneaten food – promote the growth of *F. columnare* and hence it is good management practice to ensure that no food remains after feeding koi. Generally, any stressful conditions such as poor water quality – high nitrite concentrations or low dissolved oxygen – can result in outbreaks of *F. columnare* in koi. Recent work has indicated that this bacterium is less harmful in soft water than in hard water. Nonetheless, once infection with *Flavobacterium columnare* takes place, depending on the temperature, it can spread rapidly in the affected koi, leading to mortalities.

Identification of Fin Rot

A koi with fin rot will start to develop reddening of the fins which is followed by the development of small white areas on the ends of the fins. If these white areas are not treated quickly, the infection will start to eat away the fin, and eventually the whole fin may be consumed. If this happens, the infection may spread into the body of the koi, and at this point it may prove difficult to save the infected fish. A koi with fin rot may not show any adverse behavioural signs during the early stages of infection, and so close physical inspection is essential to catch the disease early before it can become a problem.

Treatment of Fin Rot

As soon as fin rot is suspected, take the infected koi out of the pond and apply a suitable topical treatment to the affected area. This may be malachite green, followed by propolis. If the infection has become more established and the fin is actually being consumed and topical treatment has not halted this, it may be necessary to cut away the infected area of fin. Before doing this, always seek advice from a vet or koi health specialist. Having first anaesthetized the koi, cut away the fin behind the area of redness using

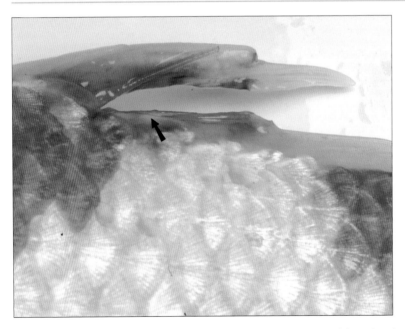
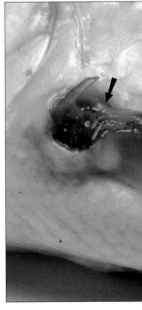

Above left and right: Advanced stages of fin rot are shown here on the dorsal fin and on the pectoral fin.

Above: Fin rot is shown here at an early stage, with the fin just starting to rot away. If caught at this stage, simple topical treatment may result in a full recovery.

a pair of sharp scissors. Once completed, apply topical treatment to the newly exposed fin to help prevent re-infection. If the fin has completely gone and the base of the infected fin is red and sore, it is possible that the infection has spread inside the koi and thus further treatment may be required and advice should be sought from your local vet or koi health specialist as to the best course of action. A swab should be done at this point to determine what infection you are treating and if the use of antibiotics is suggested by your koi specialist this will indicate which one should be used. It may also be advisable to add an anti-bacterial treatment to the pond to help lower bacterial levels.

Identification of Gill Rot

Signs of gill rot may not be noticed until the disease is quite advanced, because the gills are not easy to inspect readily. Be aware that a number of different diseases cause gill erosion, including Koi Herpesvirus (KHV). So the disease may only be suspected when it has advanced to the stage that it affects the behaviour of your koi. A koi with severe gill rot may display some or all of the following symptoms: it may be lethargic and spend much of its time on either the bottom of the pond, or at the surface in areas of heavily oxygenated water, such as around airstones and water returns. Rapid or increased gill movement may be noticed. While hanging in these areas, the affected koi may appear to be struggling to maintain its balance in the water and list from side to side. You may also notice that infected fish swim erratically from time to time and may even crash into the side of the pond. In extremely severe cases (and often these prove fatal) the eyes of the koi will appear sunken. If any of these behavioural changes are noted, it is important to catch the fish in question and inspect the condition of their gills.

Infection of the gills by *Flavobacterium columnare* occurs less frequently than on the fins and skin. Typically, infection begins at the

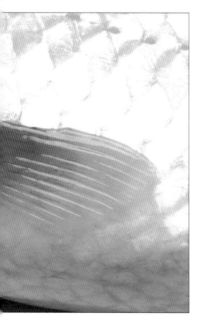

Above: Early stages of gill rot – small sections of the gill are missing.

tips of the gill filaments, progressing upwards towards the gill arch, leaving dead tissue behind the advancing bacteria. Experienced koi keepers may be able to take a small sample of the gill tissue and place this on a microscope slide, leaving it for a few minutes before examining. The *F. columnare* bacteria aggregate in strands that look rather like strands of hay in a haystack. Occasionally, infections with *F. columnare* occur extremely rapidly and become systemic, i.e. spread throughout the major organs of the body and with no immediate signs of injury to the gills.

Treatment of Gill Rot

In the early stages of gill rot caused by columnaris, using a salt bath at the rate of 100g per 4.5 litres (1 gallon) of water for ten minutes may help the affected koi. In severe cases, seek advice from your vet or local koi health specialist as to what additional treatment may be required. It may also be advisable to use an anti-bacterial treatment, such as Virkon S Aqua, to reduce the bacterial load in the pond.

Identification of Mouth Rot (Cotton Wool)

Mouth rot is included here, as it is often confused with columnaris because of the popular name, although the most likely cause is *Aeromonas hydrophila* (see page 62) At first a small white spot may appear on the nose of a koi. If not topically treated this will spread and result in an inflammation of the mouth, which in turn may result in a yellow or white discoloration of the skin. Because the skin is inflamed and may change colour, the disease is sometimes confused with fungus. A koi that is not treated will slowly lose its ability to eat and will start to look emaciated. It will also spend more time in areas of heavily oxygenated water, such as around airstones and water returns. The affected koi may also develop anti-social behaviour and spend a lot of its time apart from the other koi in the pond, appearing simply to hang in the water with its fins clamped. If not treated, the infected koi will starve because as the infection spreads it will not be able to eat.

Treatment of Mouth Rot

In the early stages simple tropical treatment with propolis may halt the progression of the disease. However, if it becomes more established it may

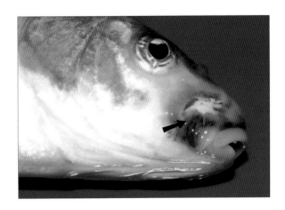

Above: This koi has mouth rot which has resulted in quite severe inflammation of the skin around the nose area.

prove necessary to seek advice from your local koi health specialist, who may advise that a swab should be taken from the mouth and sent away for analysis. The results of this determine which antibiotic to administer if antibiotics are required. This will normally only be as a last resort after other possible treatments have failed. Antibiotics will be given as an injection, as it is more than likely that the koi in question will have lost its ability to eat. It may also be advisable to add an anti-bacterial treatment to the pond to help lower bacterial levels.

Skin Columnaris Disease

Koi become susceptible to infection with *Flavobacterium columnare* for a number of reasons. Bear in mind that columnaris is associated even with healthy koi, but their immune system is able to contain any potential infection. If koi become stressed and their immune system is compromised, then columnaris quickly infects the fish. Poor water conditions, such as pollution of the pond water with either ammonia or nitrite, not only cause koi to become stressed, but also favour the spread of columnaris to all of the koi. Similarly, low dissolved oxygen concentrations in the warmest months also seem to promote outbreaks of columnaris. Poorly aerated water affects the biological filtration system, which becomes less effective, resulting in deteriorating water conditions and an accompanying increase in the organic load. The organic load in a pond is generally invisible, comprising sloughed mucus and pieces of dead cells. These form an important food resource for *F. columnare,* which then infect the koi. Once a koi has become infected

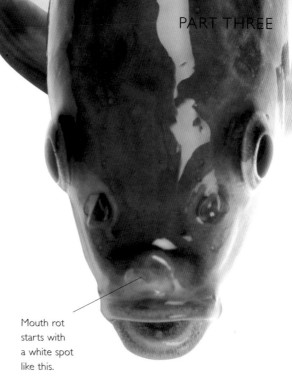

with columnaris it spreads very rapidly to become systemic, which is usually fatal.

Treatment of Skin Columnaris

In its very early stages the application of a topical treatment such as propolis, or even malachite green and propolis, may prove sufficient to prevent further infection. However, once scale lifting or loss is experienced, you should seek further advice from an expert as the use of anti-bacterials, either via injection or in the feed, may be required as a last resort. A swab should be taken and the results of analysis followed regarding the anti-bacterial to use. While bactericides are administered, topical treatment should be maintained until the koi starts to show signs of recovery. Then the frequency of this can be reduced. If, however, the infected koi is heavily swollen and many scales are lifted, recovery is unlikely and the best course of action may be to humanely kill the koi in question to prevent any prolonged suffering. If numerous koi are showing symptoms, it may be necessary to administer a pond treatment to lower bacterial levels.

Mouth rot starts with a white spot like this.

General Pond Treatment

If a columnaris infection is suspected, it is a good idea to add an anti-bacterial treatment such as Virkon S Aqua to the pond to lower the bacterial load in the water. Treatments such as Chloramine T, at a dose rate of 1g per 4500 litres (1000 gallons) can be used, but only with extreme caution in soft water, as this chemical becomes harmful to the koi under these conditions. Alternatively, potassium permanganate is preferable in soft water areas at a dose rate of 1.5g per 1000 litres (220 gallons). In addition to treating the water to reduce the bacterial load, it is important to reduce the overall organic load in the filters and pond through increased maintenance and water changing.

Above: Both koi have mouth rot. The top picture shows the early stages which can easily be treated topically. The bottom picture shows the advanced stages of the infection when specialist professional treatment is needed.

Prevention

Flavobacterium columnare particularly cause infection in koi that have become stressed, so it is good management practice to identify the underlying cause of the stress. Physical injuries that result in skin damage are also susceptible to infection, although healthy koi are unlikely to be affected as their immune system will control any bacteria that enter the wound. Low levels of dissolved oxygen in the water, overstocking, and poor water quality (especially high levels of ammonia) may make infection by columnaris more likely. Good husbandry and regular system maintenance will help in preventing an outbreak of columnaris bacteria. One factor which may make a columnaris outbreak more likely is higher water temperatures. Unfortunately this puts the koi keeper in a Catch 22 situation as many other koi diseases are less likely to prove a problem in stable warmer water. This fact has resulted in a surge in the popularity of koi pond heating. So if you do heat your pond and maintain temperatures of over 15°C (59°F) all year round, keep an eye out for any of the symptoms of columnaris infection.

COSTIA

Ichthyobodo necator is the scientific name for the parasite that most koi keepers know as costia. Costia is the smallest of the single-celled parasites that infect the body surface and gills of koi, just visible under the microscope at a magnification of 400x. Most koi probably harbour a few costia but if the health of the fish is compromised, costia can readily cause a serious infection. Costia occurs in two forms on the koi: detached or attached. Detached costia move around the body surface by means of two whip-like hairs known as flagellae. The attached form is the feeding stage; the parasite assumes a droplike shape, the 'sharp' end penetrates the skin cells and consumes the contents.

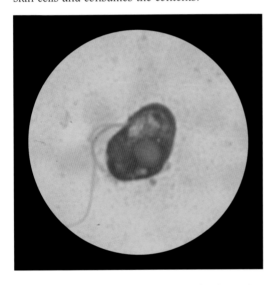

Above: It takes a well-practised eye to spot the shape of the costia parasite under the microscope.

Identification

Costia cannot be seen with the naked eye, so identification is only achieved by taking a skin scrape and viewing it under a microscope. Physical symptoms of costia are similar to other external parasites and these include excess mucus production, a common diagnostic sign of costia on koi. In severe cases reddening of the skin and open wounds may also be present. Often the head will appear to have a white film over it, and the fins of infected koi become reddened. Along with this, the infected koi may become lifeless and hang in the water while clamping their fins.

If spotted early, costia can easily be eradicated, but on young koi it will develop quickly and

can result in losses. If heavy losses are being experienced, costia should be investigated early on as one of the possible causes. In severe cases fish may also go off their food and become emaciated. Koi may also suffer from difficulties in breathing and show erratic gill movements while spending long periods of time near the surface in areas of highly oxygenated water, such as near airstones and water returns. Finally, your koi may be seen to rub and flick against the side of the pond to try and relieve the irritation caused by this parasite, which in turn can lead to secondary fungal or bacterial infections on skin abrasions.

Prevention

Costia mostly becomes a problem when koi are under stress. This is normally caused by environmental factors such as poor water quality, changes in temperature, and poor system maintenance. Be sure to carry out regular water changes, filter discharges and water testing.

Costia is active across a broad temperature range from 2°C to 30°C (36°F to 86°F), so it is important to ensure that koi remain stress free and that water conditions are monitored in the pond, even in the winter. However, a sudden temperature shock can trigger an infection with costia. When moving a koi it is sensible to ensure that the water temperature in the new location matches that of the pond from which it was moved. Similarly, if transporting koi, it is important that rapid temperature changes do not occur, say as a result of leaving bags exposed to sunlight or in the boot of a stationary vehicle. Costia can cause significant mortalities in a koi pond if unchecked.

Treatment

Outbreaks of costia tend to respond well to many of the anti-parasite remedies available from aquatic retailers and often a single treatment is enough to curb the infection with this parasite. An alternative is to use a combination of malachite green and formalin, known as the Leteux-Meyer mix, which has a more potent effect than either chemical on its own. Preparing a mix of malachite green and formalin is not for the novice; it must be used with care, as the volume of malachite green must be less than when it is administered alone as a pond treatment. Contrary to popular opinion, malachite green and formalin should be used at lower temperatures, preferably

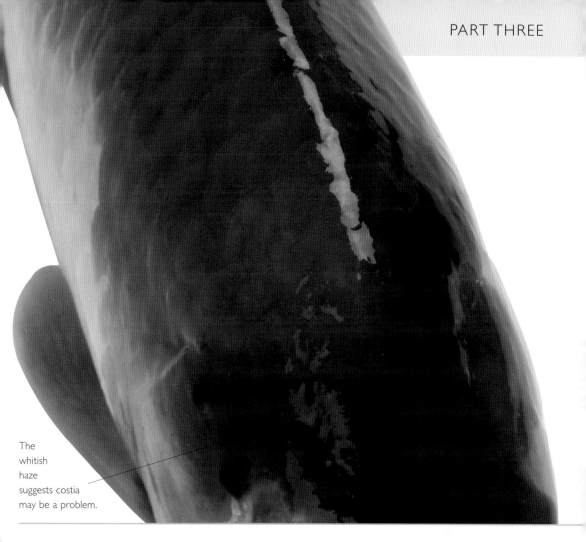

The whitish haze suggests costia may be a problem.

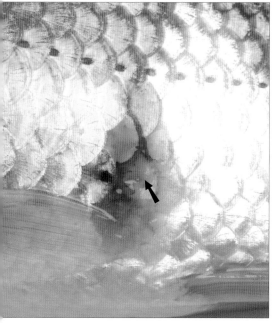

Above: A tell-tale sign that costia may be present is the presence of a milky haze over the skin, most noticeably on the head and shoulder region of the koi.

Left: As with most parasite infections, flicking will occur when costia is present, which may result in physical damage to the skin of the fish, as seen here.

below 15°C (59°F), as these chemicals become increasingly toxic as the temperature increases. Formalin also has a negative impact on the dissolved oxygen concentration, so it is essential to ensure a minimum 6mg per litre dissolved oxygen before adding a treatment based on this chemical to the koi pond. Traditionally, salt has been used as a treatment for costia, but formalin and salt are a toxic combination for koi. If a formalin-based product has been used, be sure to carry out water changes over a period of 7-10 days before using salt. It is worth mentioning that although salt has been quoted as beneficial in controlling costia, there is some evidence that this parasite is resistant to it.

CURVATURE OF THE SPINE

Koi may suffer from curvature of the spine, which may be bent laterally, termed scoliosis, or dorsally, known as lordosis. It may even be due to compression of the vertebrae. There are a number of reasons why the spine may become deformed as a consequence of both infectious and non-infectious disease. Certain bacterial infections tend to be present in those areas of the body where there are major blood vessels. The aorta runs immediately beneath the spine and bacteria may therefore penetrate the vertebrae. Recovery from such infections often results in deformity of the spine. There are even parasites, such as *Sphaerospora renicola,* which infect the kidneys and swimbladder and which in severe infections can also cause the spine to become deformed.

Non-infectious diseases that result in spinal deformities include nutritional and inherited disorders, as well as compression through large soft tissue tumours. Some fish species, such as orfe or rudd, are sensitive to many chemicals used in pond treatments and suffer distortion of the spine. Finally, exposure to electrical currents, either through earth leakage of pond equipment or as a result of an electrical storm, may cause the koi spine to break, resulting in deformities.

Identification

This is an easy condition to spot as your koi will appear as if it has a kink or bend in its spine, normally between the dorsal fin and the tail. In extreme cases it may not be limited to just one kink or bend – your koi may have a number of these creating a Z-shaped appearance.

Prevention

Usually this condition occurs in mature koi for the following reasons: overdose of medication or use of a particular medication, electric shock, lightning strike, and malnutrition. We shall consider each in turn. In the past, organophosphate insecticides were used to treat certain parasites, such as fish lice and anchor worm. Organophosphate insecticides were potent neurotoxins and susceptible species of fish, such as orfe or rudd, would suffer from bent spines after exposure to these chemicals. As well as being injurious to some fish species, the

Right: Curvature of the spine as shown here may be caused by numerous factors, such as electric shock, lightning strikes, malnutrition, or overdosing with certain medications.

organophosphate insecticides were extremely poisonous to most freshwater life and caused significant damage to the ecology of many rivers and lakes. As a consequence, these have been banned in most countries. Organophosphate treatments are being replaced with more modern medications such as emamectin benzoate, mebendazole or praziquantal, all of which are more appropriate for koi treatment. However, even some of these treatments may be potentially harmful if an accidental overdose is applied to the koi pond. It is vital to know the exact volume of your pond before adding any medication. The best way to determine this is to use a flow meter when filling the pond. You can buy or hire these from your local koi outlet.

Problems with electrical items within the pond, such as pumps or UV units, can also cause curvature of the spine, as an electric shock may affect the fishes' central nervous system resulting in kinking or bending. Always run all electrical items from an RCD (residual-current device) as this will cause them to be switched off as soon as a fault occurs. Lightning strikes may also cause curvature of the spine, causing koi to flex violently, risking spinal damage as a result.

Physical deformities such as curvature may also be caused by malnutrition and using an incorrect feed, especially in the case of newly hatched fry and juvenile koi. It is important to offer a suitable high-quality feed all times and to use mineral and vitamin additives appropriately. Bad handling techniques may also lead to a koi being dropped and injuring its spine. Certain internal bacterial infections have been linked with spinal damage, but these are relatively rare.

Treatment

Once a koi exhibits a bent or kinked spine, there is very little that can be done to correct the condition. A koi with a bent spine may live a normal life without any adverse effects except aesthetic ones. In some severe cases, however, an affected koi may start to lose the ability to swim correctly which can result in it not being able to feed. In this case, or if the affected koi appears to be in distress, it may be necessary to consider euthanasia to prevent prolonged suffering.

Right: Even a koi with quite severe curvature of the spine can continue to live a normal happy life. As there is no treatment for this condition, just leave the fish alone.

DROPSY

When encountered in a koi pond, dropsy is often confined to a single koi, but there may be a number of causes for it. Infectious disease is one, but many are non-infectious, such as the presence of tumours or major organ failure. Dropsy can, however, be induced by a viral infection, although this is seldom seen by the hobbyist, but when this does happen a number of koi will start to show the symptoms of dropsy at the same time. Dropsy is due to retention of fluid within the tissues and abdomen. It is often difficult to determine why a koi develops dropsy, but an underlying bacterial infection is the most likely to respond to treatment. Dropsy arising from non-infectious disease is unlikely to respond to most of the proprietary medications and certainly the more potent drugs, such as antibiotics, will have no effect on the condition. Often, if dropsy is due to non-infectious disease, there is little that can be done to improve the condition and the kindest thing is to euthanase the affected koi.

Identification

Early signs of dropsy include swelling of the body and protrusion of the eyes. Following these early symptoms the body continues to swell, and this results in the scales on the infected koi lifting, causing the fish to take on a pine-cone appearance. In addition, the koi is anaemic, with pale pink gills and swollen lamellae, as the blood becomes diluted with fluid. If you suspect dropsy, you may also notice a decrease in appetite plus a tendency for the infected koi to remain at the water surface and stay close to areas of high oxygen, such as water returns and outlets from the pond.

Prevention

You may have the most advanced, well maintained pond but still experience dropsy, for it is hard to completely prevent this disease. The only obvious precautions to take against dropsy are to ensure that a well balanced and healthy diet is fed at all times, water quality is good, the water is heated, and all basic husbandry procedures are maintained to the highest level. When purchasing new koi, avoid any which exhibit any of the early symptoms of dropsy. Unfortunately, even this may not prevent dropsy from occurring as sometimes it just strikes without any explanation – organ failure due to old age, for example.

Treatment

When the symptoms of dropsy are spotted, you should isolate the affected fish. However, the practicalities of doing this depend on the size of the koi, and also the size of the quarantine facilities which you have available. It is pointless isolating a 60cm (24in) koi in a 100cm (40in) tank as the stress this will cause will outweigh any advantage obtained by moving the infected koi in the first place. The first treatment for dropsy should be the introduction of salt, and this should be applied at the level of around 5 to 6kg per 1000 litres (11 to 13lb per 220 gallons) of water for at least three to five days, or until improvement is seen. However, it is sensible to isolate the affected fish in a temporary holding pond to avoid subjecting all your fish to salt and thermal stress, especially as many types of dropsy are very hard to cure. The water temperature in the treatment pond should also be slowly increased to over 25°C (77°F) and perhaps even as high as 30°C (86°F). This should be done at the rate of 1°C every day or two.

To this salt treatment you may wish to add a good anti-bacterial medication which is safe to use with salt, e.g. Acriflavine. Whatever medication is chosen, follow the directions and complete a course of treatment before reassessing the situation. One of the most distressing things about dropsy is that although it can sometimes be cured, in most cases it proves fatal. This is generally because by the time external symptoms are spotted, irreparable internal damage and/or infections have occurred, mainly to the kidneys, and they are beyond treatment. For this reason treatment should be assessed constantly and if the symptoms seem to be worsening after five days or so it may be kinder to consider euthanasia.

Below and right: Dropsy is easily spotted by the swelling of the body. In advanced stages your koi may take on a "pine-cone" appearance and have bulging eyes.

In advanced stages of bacterial dropsy reddish areas may appear on the body.

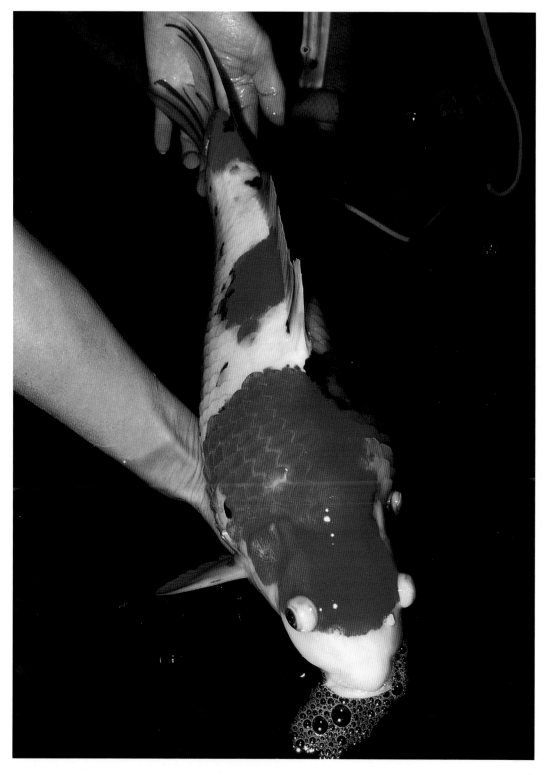

SWOLLEN ABDOMEN AND TUMOURS

Occasionally a koi may develop an unusual physical characteristic, such as a swollen abdominal area or a growth on a fin. These can be caused by a number of factors: some are hereditary, others are induced by changes in the environment or the presence of a bacterial or viral infection. If a tumour is found, nine times out of ten it will prove to be harmless. If it is internal, little can be done other than careful monitoring of the affected koi. If the growth is external, it may be possible for a vet to remove it surgically but the chances are that it may return once removed, and there is also a risk that the area from which it is removed may becoming infected. So generally the best approach is to leave suspected tumours alone and just monitor the affected fish. The only time action should be considered is if the koi appears in distress and loses its ability to eat. Then euthanasia should be considered as the humane option.

If the swelling is in the abdominal area of a female koi, it is possible that the fish is egg-bound. Other possibilities – although they are very unlikely – include the presence of worms or the initial stages of dropsy. Some internal growths and egg retention can put pressure on the swimbladder which results in the koi losing its ability to stay upright in the water.

Identification

If you suspect that a koi is egg-bound, first confirm that the affected fish is female. A koi needs to be around 30cm (12in) or longer to sex. It is possible to identify a female koi by comparing body shape and fin size – female koi tend to be broader across the shoulders and fuller bodied with a cigar shape, while male koi tend to be more slender with a torpedo

Below: This koi has an internal tumour. As it is a female, it may possibly be of the ovaries, although only a post-mortem will allow an exact identification to be made.

Above: This koi shows a swollen and rather uneven abdomen which, coupled with slightly raised scales, generally indicate the presence of an internal tumour.

Below: Tumours or cysts commonly affect female carp and koi. These large growths were found in the ovaries of this koi, which caused the abdomen to become distended.

shape, and the fins on a male koi may be larger and less rounded at the ends. However, the most reliable method of sexing is to inspect the vent area to determine if the koi is female or not. A female koi will have a line running from head to tail in the vent area crossed at one end with another line running from side to side to give the appearance of a T. A male koi will simply have a line running from head to tail. Having determined that the koi is female, gently feel the swollen area; if it is soft to the touch but is not too fluid, it could possibly be unreleased eggs. If the area feels hard, it could still be eggs but is more likely to be a tumour. However, without invasive surgery,

Determining The Sex Of A Fish

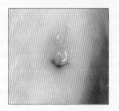 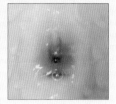

Male	Female

One way to sex a koi is to examine the shape of the vent. Females have a line running from head to tail and another line across it at the tail end. Males just have a simple line running from head to tail. However, this difference is not easy for a novice to distinguish – you need an experienced eye. Males also develop a spawning rash – fine raised pimples around the head, top of the body and on the bony rays of the pectoral fins.

which is normally ineffective, it is impossible to make an exact diagnosis.

Prevention

Just because you have a pond containing male and female koi, it does not mean that the females will spawn. Temperature and day-length are the critical factors that will bring the koi into good spawning condition and allow the eggs to mature fully in the females. Eggs are generally released during spawning, which takes place when the water temperature hits 20°C (68°F) or higher. It is a popular misconception that female koi resorb their eggs during the winter as a source of nutrients. Koi that are well fed and in good condition do not need to use eggs as a source of nutrition. If a female koi is seriously malnourished or very sick and has not been feeding for a period of weeks, the eggs are used as a food source, but these are rare or unusual situations. Unless you are a koi breeder, it is best to maintain a constant temperature throughout the winter of around 16°C (61°C). However, if you have female koi which are to be used for spawning, these should be allowed a colder spell for around two months. This lets these koi use up excess reserves, i.e. unused

Left: This koi has a suspected tumour of the reproductive organs. In some cases it can be difficult to determine if a koi has a tumour or another condition such as dropsy, which can cause similar physical symptoms.

eggs, and get into condition for the following season, making the chances of a koi becoming egg-bound less likely.

Treatment

It can be very difficult to determine whether a koi has an internal tumour. Often the first indication is an unusual bulge or lump on the side of the fish. In many instances the bulge is only just noticeable to an experienced eye, but it is worth stating that these fibrous types of cyst or tumour are very common in female carp and therefore koi. Unfortunately, there is little research into the origins of these tumours or cysts and the underlying nature of these lumps is largely unknown. In terms of aquaculture, carp are of marketable size for the table before these fibrous cysts and lumps develop, so there is little incentive to investigate them further. Usually the fibrous lumps appear to cause no distress to the koi and they feed and live for many years despite these growths. Sometimes the lumps do increase in size and can cause serious health problems. They may compress the swimbladder, leading to buoyancy issues, or press soft tissues and blood vessels, which in turn affects the normal physiology of the koi. At this advanced stage of the cyst or tumour growth, the koi is generally showing signs of distress and will cease feeding.

Regrettably, there is no treatment for these tumours or cysts. Although in the early stages koi may show no signs of discomfort or distress, it is important to understand that there will come a time when the growth will impact on the koi's health. As it becomes debilitated by the growth, in most instances it will no longer swim with the others but spend increasing amounts of time either resting on the bottom or near the water or air inlet. If buoyancy is affected, the koi may struggle to reach the surface and swimming through the water may be accompanied by disorientated spiralling motions. At an advanced stage, the koi stops feeding. Taken individually or together, these are signs that the koi is in distress and, as painful as the task may be, the kindest option is to euthanase the fish. It is worth seeking professional help with euthanasia as failed attempts to cull a koi are worse than not intervening at all.

EPISTYLIS

Epistylis is a ciliate protozoan invisible to the naked eye, so exact identification must be via a skin scrape examined under a microscope. When viewed like this *Epistylis* looks bell-shaped with a long "handle" connected to it. Tiny hair-like cilia on the end of the bell shape may be seen. Epistylis secretes a small pad of a glue-like substance that holds it in place. The cilia around the top waft bacteria and particulate organic debris into the mouth. *Epistylis* may also be seen in its contracted form and in this instance it will simply look circular.

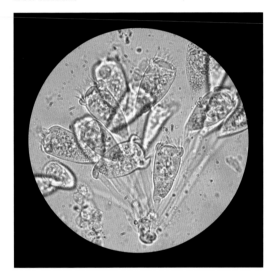

Above: The long bell-like shape of *Epistylis* is visible under the microscope, but it takes a practised eye to make an accurate identification.

Identification

In the early stages of infection no external visible signs may be seen on the infected koi, but behavioural changes such as flicking or hanging in the water may be observed. At this point microscope examination of a skin scrape is the only way to make an exact diagnosis. As the infection worsens, small white patches may appear on the skin which can develop in size up to 5mm (0.2in) or so. At first these may be limited in number, but as they become larger they will spread and more will become apparent on the body of the infected koi, as well as on the gills. As these small white patches spread, the skin will become redder and this eventually leads to scales lifting and (if not treated) falling out. This leaves the affected area susceptible to secondary

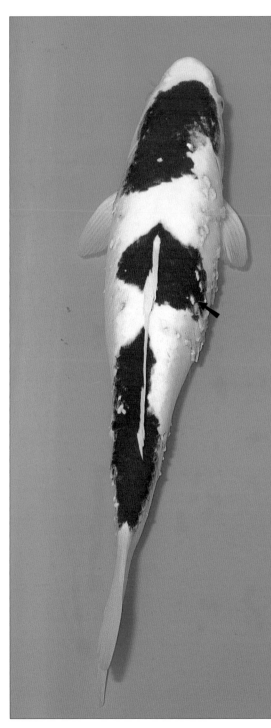

Above: This koi is in the later stages of an epistylis infection. The white patches on the skin are turning into areas of redness, and scales may be lost as the infection develops.

Above: This fish exhibits the early stages of epistylis. If treated at this stage, the loss of scales may be prevented.

infections from bacteria or fungus.

At this stage of infection it is vital to check regularly for other parasites as these will quickly take advantage of the situation and worsen the problem. As the epistylis infection spreads, it is quite common to see ulceration caused by bacteria attacking areas from where scales have fallen. The glue-like substance that holds the *Epistylis* in place is an irritant and, if present in large numbers, the koi will repeatedly flick and rub, damaging the skin and the ability to produce the protective mucus. If left unremedied, the koi will develop secondary bacterial and parasitic disease that will result in mortalities. *Epistylis* thrives in ponds where there is a rich particulate organic content and is therefore a commensal, rather than a parasite. Infections occur at all temperatures, but *Epistylis* will reproduce more quickly and increase the number of colonies on the koi as water temperatures rise.

Prevention

As with most diseases improved husbandry and system maintenance can reduce the likelihood of infection and epistylis is no exception to this. Outbreaks of *Epistylis* are never stress related, but simply due to poor husbandry, as the particulate organic material in the pond

has been allowed to increase, encouraging a high bacterial load on which this commensal is feeding. As water temperatures rise, the bacteria will reproduce faster and provide more available food for the *Epistylis*, which will increase in numbers. If spotted early, *Epistylis* should not cause a major problem, but if allowed to get to an advanced stage it will leave the infected koi highly susceptible to other infections. If *Epistylis* is experienced, take extra care to check for the presence of any other infections which might move in and create a whole new set of problems.

Treatment

Epistylis can be treated with an off-the-shelf parasite treatment, or alternatively with malachite green alone. Dose rates will depend upon the concentration of the solution. Alternatively salt can also be used to control epistylis effectively. It should be used as a bath, at the dose of 100g of salt per 4.5 litres (3.5oz per gallon) for 10 minutes, which should be accurately timed. This can be repeated for three consecutive days. Unless the infection is halted early, secondary infection may well occur and this will generally need specific treatment. This can be done via topical treatment with either propolis alone, or malachite green and propolis in more severe cases. Fungus is also common and in most instances this will need to be treated topically with malachite green and/or propolis.

FUNGAL DISEASE (COTTON WOOL)

Fungal disease in a pond may be caused by a number of fungi which are naturally present in most bodies of water, for example the aquatic fungus *Saprolegnia*. These fungi will not prove a problem to healthy koi in a well maintained pond, and in fact their presence helps to maintain a balanced ecosystem as they live on decomposing and decaying matter. These fungi will only attack damaged areas of skin or decomposing waste. So fungal disease is a secondary infection as it requires a koi to be already damaged, for example by a parasite. Only then will fungus attack the weakened area of skin, fin or mouth. Fungus will also infect eggs that have not been fertilized or where the embryo has died. These eggs tend to be white or appear cloudy rather than have the transparent appearance of healthy eggs. In a hatchery environment, these dead eggs would be removed with a pipette, but in the pond it is most likely the koi and other fish will readily consume them before they become affected by fungus.

Identification

Developed fungal infections are easily visible to the naked eye, as the infected area will have a white tufty growth covering it. In some instances this may resemble cotton wool, though it is not always white – instead it may have a brown, black, or grey appearance. In early stages of fungal infections it is not always possible to spot an attack without taking the koi out of the pond

and inspecting susceptible areas. To begin with only one or two small white strands may be noticed, but these build up over time to become the cotton-wool-like tufts typical of a fungus infection. It is important that fungal infections are spotted early as this both enables the infection to be removed quite easily, and it also reduces the chances of death occurring from the fungus attack. In a severe fungus infection, when large areas of the body are covered with a cotton-wool-like fur, the chances of survival are dramatically reduced as the fungus will start to attack the koi internally, which leads to irreparable damage to the internal organs.

Be aware that there is a condition known as mouth rot or cotton wool mouth which looks like a fungus infection. In fact it is not caused by fungus but by the bacterium *Flavobacterium columnare*, and so it requires a different course of treatment.

Prevention

Fungus is a secondary infection and will only become a problem after something has damaged the protective skin of your koi. This could be due to a parasite infection, bacterial infection, netting damage, poor water quality, spawning or poor diet. In fact any stress factor could leave your koi susceptible to a parasite infection, for example, and thus potentially a secondary fungal infection. So the secret to preventing fungus is to provide a stress-free environment for your koi. This can be achieved by regular system maintenance, water

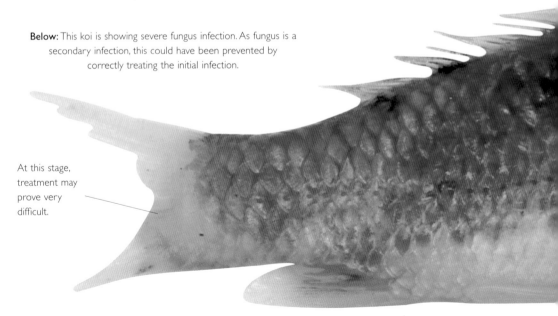

Below: This koi is showing severe fungus infection. As fungus is a secondary infection, this could have been prevented by correctly treating the initial infection.

At this stage, treatment may prove very difficult.

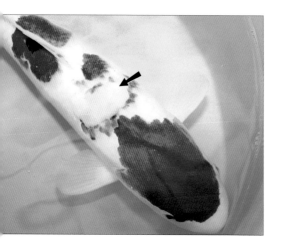

Above: This koi has an area of fungus in front of the dorsal fin. The most probable cause of this is physical damage which has gone unnoticed, and caused the fish to succumb to a secondary fungus infection.

testing, and pond heating to prevent fluctuations in water temperature.

Treatment

Before attempting to treat fungus you must identify what has allowed the fungus to become a problem and treat this first. If this is not done, you will be fighting a losing battle in trying to get rid of the fungus. A large number of proprietary fungal remedies are available which can be used as a bath or pond treatment with differing levels of success. The best approach in the case of fungal infections is to topically treat infected koi individually. If you are dealing with only a light infection, this can be done in the pond. However, in severe cases where large areas of the koi are covered, it may be preferable to treat the koi in a quarantine tank. This will help reduce the number of fungus spores that are released into the water to a minimum.

Topical treatment should be applied to the areas affected with fungus, and it may be easier to do this if the koi is sedated first. Alternatively the koi can be held while being treated. Before applying topical treatment the fungus should be removed with a cotton-wool bud, then malachite green should be applied to the area, before the whole area is finally sprayed with propolis. Then the treated koi can be returned to the pond. If, when the fungus is removed, it looks like a bacterial infection has occurred, i.e. there are large areas of ulceration, it may be necessary to take additional steps and these should be discussed with your local koi health specialist. Fungus will attack dead eggs, so if you are spawning your koi, it is a good precaution to apply a suitable anti-fungal remedy to the pond once spawning has finished, e.g. methylene blue. However, take care to maintain optimum conditions when using methylene blue as it can destroy the nitrifying bacteria in your filter. Common salt (sodium chloride) as a pond treatment is excellent for controlling fungus, which cannot grow in a brackish environment.

GAS BUBBLE DISEASE

This condition is very similar to the "bends" which divers experience when they rise to the surface too quickly – the rapid change in pressure results in tiny bubbles forming in the blood. In a koi pond gas bubble disease mostly occurs as a consequence of supersaturation of the water with dissolved nitrogen, as 72% of atmospheric air comprises nitrogen gas. Supersaturation occurs for a number of reasons, and is generally encountered at lower water temperatures when water will hold higher levels of air. Dense algae

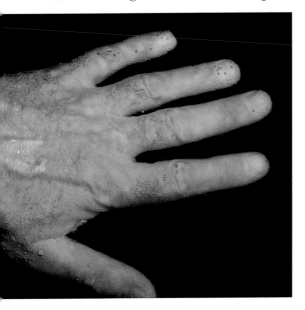

Above: One sign that supersaturation is occurring is the formation of tiny bubbles on the pond wall. This condition is illustrated here by the presence of very small bubbles coating the surface of a hand immersed in the water.

blooms may cause high dissolved oxygen concentration leading to gas bubble disease, but such heavy phytoplankton blooms are more characteristic of eutrophic natural ponds than a koi pond. A pinprick hole in pipework or poorly sealed pipe joints can suck in air and force it into solution under pressure, known as air entrainment. In cold water, air entrainment can lead to supersaturation of water with air and koi can develop gas bubble disease. Heating cold water very rapidly maintains the air in solution and may also cause gas bubble disease. Less commonly for the koi keeper, borehole water may contain water supersaturated with nitrogen. Finally, in koi ponds, the use of pressurized

equipment such as sand filters may result in any trapped air in the unit being forced into solution in the water, causing supersaturation. Gas bubble disease can be fatal, but if the signs are spotted soon enough steps can be taken to alleviate the problem.

Identification

As a koi takes in water over its gills, the gases it contains are diffused into the blood. As water supersaturated with nitrogen passes over the gills, the nitrogen gas readily dissolves into the blood because the concentration of this gas in the koi's body is lower than in the surrounding water. The pressure in the blood of the koi is less than in the water, so the nitrogen gas drops out of solution in the blood vessels and tissues. As gas is less dense than the blood and body fluids, it forms bubbles in those areas near the surface of the body, such as the fins, eyes or even in the gills. These bubbles are known as embolisms and it is these that pose the greatest threat to koi. Nitrogen bubbles in the blood can block blood vessels and affect the circulation. In severe cases they may result in abnormal behaviour, unstable swimming and even death. Oxygen supersaturation caused

Below: Gas bubble disease is very seldom experienced and can be very hard to identify. Here the presence of small bubbles in a fish's dorsal fin can just be seen.

Above: The pectoral fin of a koi suffering from gas bubble disease. Here the bubbles can be seen radiating out along the fin away from the body of the koi.

by dense algae blooms will cause the same problems for the koi as excessive dissolved nitrogen. External signs of gas bubble disease include small gas bubbles that are visible on the fins, around the eyes, and on the gills. Nitrogen bubbles may even cause the eyes to swell and pop-eye to develop. If air bubbles become trapped between the two outer layers of skin, you may notice that the skin starts to lift, and as more bubbles collect in this area the more pronounced the lifting will appear.

Prevention
Gas bubble disease occurs when the water becomes supersaturated, so the best prevention is to stop this from occurring. Air entrainment through faults in the pipework and during periods of cold weather is probably one of the commonest causes of gas bubble disease. It is therefore a good idea to check all joints and pipes for any slight leak or fault each autumn and before cold weather. If your koi pond is heated, make sure the system does not heat the water too rapidly. This is best tested before the cold weather. These days, dense algae blooms are a rare occurrence in the koi pond due to better filtration systems and UV clarifiers. Gas bubble disease can also

occur when gases become trapped in pressurized filtration equipment, such as sand filters, so it is a good idea to bleed off any excess air build-up in the unit when using such items. Good aeration also reduces the chances of this disease occurring as it will dispel the excess gas from the water.

Treatment
The first thing you should do is start to exchange large amounts of the water for water which is not supersaturated as this will help bring the levels of dissolved gases back to normal. When doing this ensure that dechlorinated and conditioned water is used to avoid causing excessive levels of stress. You should also try to create as much surface disturbance as possible as this will encourage the excess gas levels in the water to diffuse back into the air. Normally if these steps are taken, no further treatment will be needed and the koi should return to health. If, however, nitrogen bubbles have formed in the blood, there is a chance that blood vessels may have become blocked, and so sudden losses may still occur.

GILL DISORDERS (STICKY GILLS, GILL MAGGOTS AND GILL ROT)

Gill problems can be caused by numerous factors including parasites, many of which are described elsewhere in this section. Other factors which can cause gill problems are gill maggots, poor water quality which can result in sticky gills, and the fungus *Branchiomyces* which will cause gill rot. Bacterial disorders and the viral disease Koi Herpesvirus (KHV) also result in gill damage and these pathogens are discussed elsewhere in this section.

The problem with gill diseases is the fact that they are not easy to spot in the early stages of infection because the gills are covered by the operculum. The problem is only normally apparent once your koi show other symptoms, such as gasping, hanging in areas of heavily oxygenated water, such as waterfalls and airstones, and even going off their food. When these symptoms are noticed, the koi in question should be taken out of the pond, and the gill cover gently lifted for closer examination of the gill. The problem with treating diseases of the gills is that it depends on the cause of the complaint. While many bacterial or parasitic diseases will respond to treatment, there is no cure for any viral diseases, which includes KHV. Often, despite treatment, koi losses may continue to be experienced, either due to secondary infections or simply because the koi has become weak and debilitated as a result of the gill disorder.

Above and below: Here are two tell-tale signs that a potential gill problem may be occurring. The top image shows very pale gills typical of anaemia, while the bottom picture shows gill filaments which are sticky.

Sticky Gills

Lamellar hyperplasia, or sticky gills, is normally caused by poor water quality or low oxygen levels resulting from lack of aeration. As the name suggests,

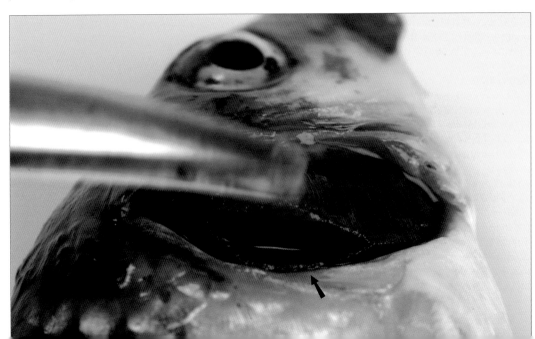

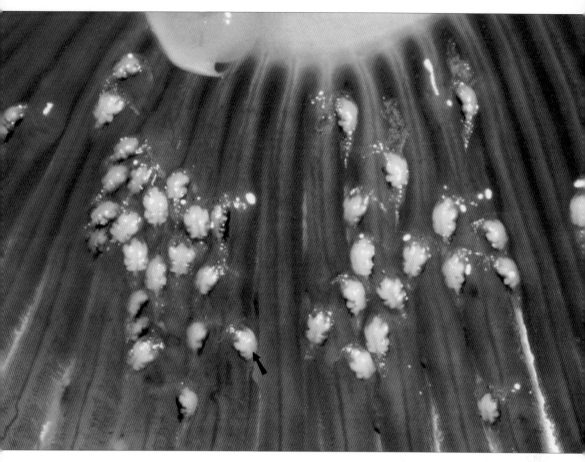

Above: A section of a gill which is infected with gill maggots. This is seldom experienced by the hobbyist; however, if you think that this may be a problem, seek specialist advice.

this complaint is easily identified because the gill filaments will be clumped together, and may appear to be covered in mucus. The first thing to do is test the water quality, as pollutants such as ammonia cause sticky gills. If the water quality is poor and there is a high concentration of ammonia make small (10 – 20 per cent) water changes with dechlorinated water and increase aeration. It is important to check for the presence of other diseases. An indication that these might be present will be if any of the gills are grey in colour or if any of the filaments look as if they are being eaten away or rotting. If any of these signs are spotted, it may signify the start of something more serious so a skin scrape should be taken from the koi in question starting as close to the gills as possible. If a bacterial infection is suspected, a swab should also be taken from around the gills. Having ruled out the presence of anything more sinister, and if water quality tests have identified the

underlying problem, the best course of action is to give the infected koi a salt bath once a day for three consecutive days. This should be done at a dose rate of 100g of salt per 4.5 litres (3.5oz of salt per gallon) for ten minutes, remembering to add plenty of aeration to the water being used in the bath. This, combined with correction of the water quality in the pond, should lead to a rapid improvement and full recovery, as this is not a particularly serious condition.

Gill Maggots (Ergasilus)

Gill maggots are parasitic crustaceans which, if left untreated, can cause major gill damage, and result in large fish losses. These parasites are found on the gills, gill covers, and the mouths of infected koi. Only the female gill maggot is parasitic and, as the name implies, is found on the gills of fish, where she feeds on the tissues, eroding away the structure of the delicate gills.

This is replaced by fibrous scar tissue, which then compromises breathing and gas exchange. During the spring and summer the female produces two egg sacs that are shed into the water. The eggs hatch to give rise to free-living copepodid stages, during which time the juveniles go through periods of growth followed by moulting. On reaching sexual maturity, the gill maggots mate, the males die and the female seeks a fish host. The females overwinter feeding on the fish host and begin to produce eggs in the following spring.

Gill maggots are more common amongst wild fish and are generally rare in the koi pond environment, or indeed the ornamental fish industry. Infection takes place through the introduction of fish carrying female gill maggots.

Because of the danger of introducing gill maggots into the koi pond, mixing koi with wild fish is a poor management practice. In the past, infections with parasites such as gill maggots have been difficult to eradicate because of the lack of suitable treatments that would not also impact on other freshwater life. In recent years, the introduction of medications incorporating emamectin benzoate have proven to be an acceptable method of controlling crustacean parasites in the ornamental fish sector.

Branchiomyces – **Gill Rot**

This condition is generally only a problem for koi that are housed in poorly maintained ponds, especially those with a high particulate organic

The life cycle of the gill maggot (Ergasilus)

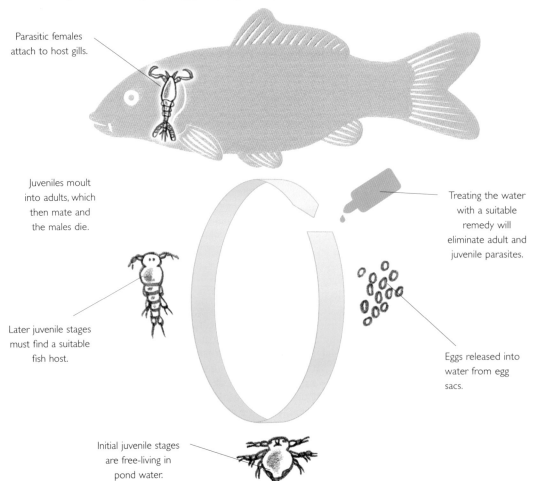

Parasitic females attach to host gills.

Juveniles moult into adults, which then mate and the males die.

Later juvenile stages must find a suitable fish host.

Treating the water with a suitable remedy will eliminate adult and juvenile parasites.

Eggs released into water from egg sacs.

Initial juvenile stages are free-living in pond water.

Above: In the more advanced stages of gill disorders, the actual filaments of the gills may start to rot away or physically disintegrate, as seen here.

load. There are two species of *Branchiomyces*: *B. demigrans* affects both the gills and the body and is frequently confused with *Saprolegnia*. The second, *B. sanguinis*, can only grow in tissues with high oxygen concentration, so is commonly found infecting the gill capillaries. Koi infected with this fungus start to show signs of respiratory distress, hyperventilating (often termed 'puffing') and will move to well-aerated areas of the pond. Examination of the gills shows them to be discoloured, with a light and dark mottled appearance due to pinprick haemorrhaging, clotted blood and areas of dead tissue. There is also lamellar hyperplasia and fusion of the gill filaments. The only way to identify this fungal problem is to examine some of the infected gill filaments under a microscope. Because of the dangers involved when taking samples from the gills, this procedure should only be attempted by a suitably qualified person. Even with a gill

sample, it may prove incredibly difficult to make an exact identification.

It can prove difficult to control infections of *Branchiomyces*. The best approach is prevention, ensuring that the particulate organic load is kept to a minimum and controlling dense algae blooms. If a suspected outbreak does occur, it is best to confirm that the gills are infected with *Branchiomyces*, while at the same time improving the environment in which the koi are housed by regular water changes. Salt can be used as a bath at 85 to 100g per 4.5 litres (3 to 3.5oz per gallon) for ten minutes once a day for three consecutive days. Remember to provide added aeration in the bath.

Alternatively a proprietary gill medication may be applied to the pond, or treatment with Chloramine T could be considered at a minimum dose of 1g per 1000 litres (220 gallons), but use it with caution in soft water areas. The problem with this disease is that not only is it hard to eradicate, but it also causes physical damage, from which the gill may not regenerate. Consequently losses may occur for some time.

GYRODACTYLUS (SKIN FLUKES)
AND DACTYLOGYRUS (GILL FLUKES)

These are the two most common types of worm infection which will be encountered by the koi keeper. If identified early, they can normally be easily eradicated. Both *Gyrodactylus* spp. and *Dactylogyrus* spp. are external parasites which secure themselves to koi by means of their specialist attachment organs. *Gyrodactylus* – more commonly known as skin flukes – are worm-shaped and have a set of hooks for fastening on to a koi located at the rear of their bodies. When viewing skin flukes through a microscope (they are not visible to the naked eye), up to four developing flukes can be seen within the adult, each one located within the next. Skin flukes give birth to live young, but only one at a time. The reproduction rate depends on temperature, and infections are common in the spring. Once a skin fluke is attached to the host koi, it will live on mucus and skin. Skin flukes are passed by contact between koi, and healthy fish may harbour numbers of them without any ill effect.

Dactylogyrus – or gill flukes – infect the gills of koi. They also have hooks located at the rear of their bodies to attach to the host koi, and, as with skin flukes, these are surrounded by a number of smaller hooks. Unlike skin flukes, gill flukes do not bear live young; in fact they are hermaphrodites, so each worm can produce and fertilize a single egg at a time. Once the egg is laid it is shed into the water irrigating the gills. Hatching would seem to be stimulated by the presence of a passing koi. The freshly hatched gill fluke is attracted to the koi, probably by a chemical stimulus, and can swim actively to seek a host by means of cilia, or tiny hairs, that surround it. Water temperature dictates the rate of growth and reproduction. Unfortunately, the confined environment of the koi pond favours the parasite in its ability to find another host and start reproducing.

Below: A gill fluke seen under a microscope. A skin scrape is essential for making a definite identification of flukes. Try to take the mucus sample as close to the gills as possible.

The life cycle of the gill fluke (Dactylogyrus)

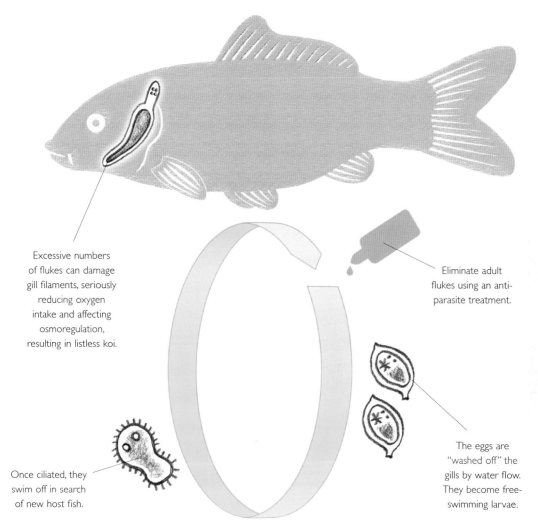

Excessive numbers of flukes can damage gill filaments, seriously reducing oxygen intake and affecting osmoregulation, resulting in listless koi.

Eliminate adult flukes using an anti-parasite treatment.

The eggs are "washed off" the gills by water flow. They become free-swimming larvae.

Once ciliated, they swim off in search of new host fish.

Identification

Neither *Gyrodactylus* nor *Dactylogyrus* is visible to the naked eye and thus a skin scrape must be viewed under a microscope for a positive identification. A skin scrape from the body is sufficient to identify both types of flukes, as gill flukes will normally attach themselves to the body of a koi and make their way towards the gills over a number of days. Both types of worm may appear to move over the slide by a process of expansion and contraction. When a koi is infected with skin flukes, it may start to produce large amounts of mucus, and if the infection

becomes severe, secondary infections, such as fungus or bacterial problems, may arise.

If gill flukes are the problem, the gills show characteristic areas of raised white tissue where the hooks are embedded in the filaments. The koi may produce more mucus around the gills in response to heavy gill fluke infections and both will contribute to causing respiratory embarrassment. A koi may show signs of irritation, repeatedly rubbing its head on the pond surfaces and fixtures. Affected koi may also hang in the water, close to heavily aerated water, such as near airstones and water returns. You may also

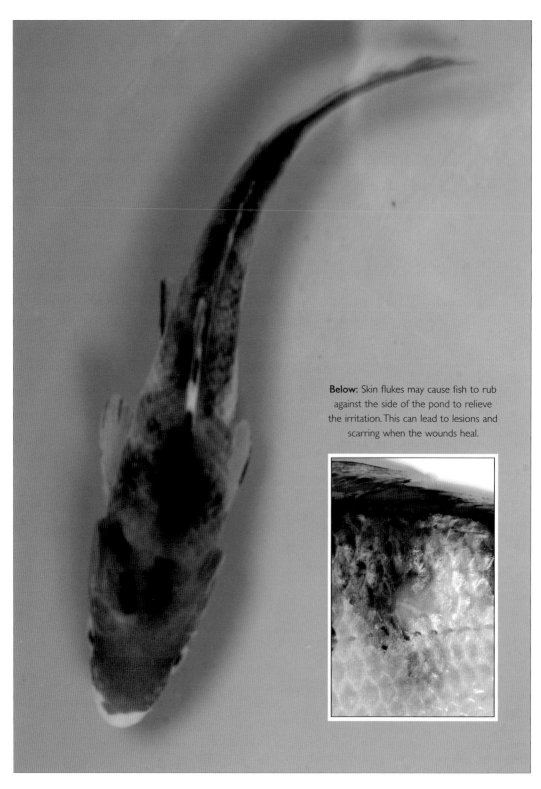

Below: Skin flukes may cause fish to rub against the side of the pond to relieve the irritation. This can lead to lesions and scarring when the wounds heal.

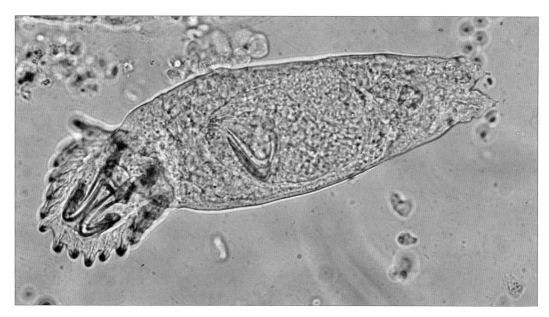

Above: A skin fluke is shown. Closer examination shows that inside the adult fluke a baby fluke is developing.
Left: A small koi with a heavy fluke infestation. As the infection takes hold, your fish may become emaciated and exhibit a milky colour due to excess mucus production.

notice that food is not taken as readily as normal.

If skin and gill flukes are not identified early on, secondary problems will occur. In the case of gill flukes if sufficient damage is done to the gills before the disease is spotted, losses may occur.

Prevention

You can expect to find the odd fluke on a koi, but it is when the koi environment deteriorates significantly that flukes start to reproduce at a fast enough rate to cause a problem. It is therefore important to maintain good water quality and to be thorough about pond hygiene, cleaning mechanical filtration and reducing organic debris to a minimum. Also maintain optimum oxygen levels and carry out regular system maintenance, including regular small water changes, to help to prevent these diseases from occurring. When buying koi, be sure to avoid any fish showing signs of excess mucus production and irregular gill movement.

Treatment

Skin flukes normally only require one treatment, as they cannot survive in the absence of a host. Gill flukes may need a course of treatments because of the period over which the eggs hatch and develop. The eggs are impervious to chemical attack. There are numerous proprietary medications designed to treat flukes, all with varying results. An alternative is to use malachite green and formalin at a dose rate that will depend on the strength of the solution. If opting to use this, ensure that the water temperature is above 11°C (52°F). If water temperatures are lower than this but still above 7°C (45°F), it may be worth considering potassium permanganate at a rate of 2g per 1000 litres (220 gallons). This can be repeated, if required, every five to seven days for three weeks, but use with caution in hard water.

In recent years there have been significant advances in the treatment of flukes within the ornamental industry. New medications based on the drugs praziquantal and mebendazole have been formulated for the pond environment and are extremely efficient at controlling fluke infections. These products are currently available only through a vet and not over the counter at retailers. In parallel with treating and eradicating the fluke infestation, it may also be necessary to treat any secondary infections. This can normally be done by applying propolis topically in combination with an appropriate bactericide.

HI-KUI

Hi-kui only attacks the skin of koi with red pigmentation, especially Go Sanke varieties, i.e. Kohaku, Showa and Sanke. The term Hi-kui is composed of two elements – hi means red and kui translates as eaten, so Hi-kui can be described as a disease that eats the red areas of pigmentation on koi. Hi-kui is widely used to describe a range of conditions, although the actual cause could be a number of things. Hi-kui may be used to describe complaints such as localized, minor skin conditions, which can normally be easily treated using the methods described below.

Overall, skin diseases in fish have received little attention in terms of their origins and pathology and Hi-kui in koi is no exception. Since it may also include malignant growth, or cancer, always seek specialist advice if in doubt. While Hi-kui appears to affect only good-quality koi and probably devalues the fish, the condition can occur on the red pigment of any koi.

Identification

Hi-kui disease may show itself in a number of ways. First you may notice a small discoloration of an area of red pigmentation, causing the red pigment to look faded compared to the rest of the red on the koi. This area may also look matt and sunken and lack the same sheen as the rest of the fish. Secondly an area of red pigmentation may take on a brown haze, and look raised compared to the surrounding area. Thirdly small dark brown areas may appear with a diameter ranging from a

pinhead up to the size of a small coin.

Whatever symptoms your koi exhibit, the problem is caused by a thickening of the overlying epidermal tissue which may be triggered by a number of factors. While Japanese koi breeders have implicated exposure to sunlight as the cause of Hi-kui, clearly a number of different skin conditions with unknown causes have been included within this blanket term. The effect of sunlight in inducing Hi-kui may well be correct, but there are other predisposing factors associated with the red pigmentation. Poor husbandry and system maintenance may make attacks of Hi-kui worse, as dirt and waste within the system can attract bacteria that view these affected areas as an ideal site for secondary infection.

Prevention

As overexposure to sunlight has been implicated as a cause of Hi-kui disease, shading the pond is a good idea. For this reason many keepers believe that koi kept in green water are less likely to get Hi-kui, although this is not an attractive solution for the typical hobbyist who wants to view their koi in a clear water. Good system maintenance and regular discharges of collected

Right: This fish is showing advanced stages of Hi-kui – the red pigmentation and skin tissue is breaking down.
Below: Hi-kui disease generally only affects higher grade koi wlth red pigmentation, such as Go Sanke varieties.

waste will all help. They do not necessarily stop Hi-kui from occurring, but they do help to prevent other infections taking hold. It is also essential to ensure that optimum oxygen levels are maintained at all times, especially in high temperatures which usually coincide with high exposure to sunlight. Despite these precautions it may be impossible to stop an outbreak of Hi-kui as it can simply appear through no fault of your own. Unfortunately it tends to occur more readily on high quality koi with strong and stable hi (red), and if your collection contains fish which fall into this category, be warned that it might just happen!

Treatment

Hi-kui is not contagious and in real terms it has no effect on the overall health of the koi, as long as the system in which they are kept is of a suitable standard and is well maintained. The main effect of Hi-kui is cosmetic – it degrades the appearance of your fish, and in the case of high value koi, it may devalue them. Whether you decide to treat or not is a matter of personal choice. If you do decide to seek treatment, the koi in question will need to be sedated before any surgical treatment can be carried out. Having sedated the fish, a vet or koi health specialist will scrape away the area of Hi-kui with a clean sterilized scalpel until the raised parts of Hi-kui are gone, or at least reduced in size if a large area is affected. Then a suitable topical treatment may be applied to the area to stop secondary infections, such as fungus. Propolis is a good treatment to apply although malachite green or similar will do.

The koi can then be returned to the pond, but the treated area of skin must be monitored on a regular basis to ensure that no secondary infections occur. It may be necessary to repeat the topical treatment on a regular basis until the area is healed. It is sensible to use an antibacterial medication, such as Virkon S Aqua, in conjunction with any other treatment to reduce the likelihood of secondary infections. Regrettably, as the underlying pathology of Hi-kui is poorly understood, it is difficult to be certain that any treatment will be successful in the long term and the condition will probably recur.

KOI HERPESVIRUS (KHV) AND SPRING VIRAEMIA OF CARP (SVC)

KHV is of increasing concern, not just to koi keepers but to producers of carp worldwide. Indeed, it has been made a notifiable disease by the Organization Internationale Epizootique (OIE), an animal equivalent of the World Health Organisation (WHO). This means there is a legal requirement to alert the authorities if an outbreak is suspected. Koi with KHV may have pathologies that are similar to many other diseases, such as bacterial gill disease or other infections that are bacterial in origin. Generally, the signs of KHV are chronic erosion of the gills, accompanied by localized hyperplasia of the skin, producing a pimply or heavy mucus effect. The eyes are sunken into the head (endophthalmia). KHV is a virus and although a vaccine has been developed to protect koi from it, at present there is no cure. Only preventative measures can be suggested.

Koi that have been exposed to the koi herpesvirus may well die, and hobbyists can anticipate losses of over 80 per cent of stock, especially if an exact diagnosis of KHV is not made early on. Koi herpesvirus is detected using either Enzyme Linked Immunosorbent Assay (ELISA) or Polymerase Chain Reaction (PCR). For the ELISA test, a small blood sample is all that is required. It reacts with antibodies produced by the koi immune system in response to KHV. The koi may not be showing any signs of infection. The positive response to the ELISA test implies a koi has been exposed to KHV, but not necessarily that it is infected. For example, koi vaccinated against KHV will test positive. The PCR test requires the sacrifice of the koi to sample the gills and other internal tissues. It is a technique used to isolate the virus in koi that already sick, where KHV is the suspected cause.

The virus is affected by temperature, so keeping your koi at certain temperatures can limit its effect. In low water temperatures (around 7°C/mid-40s°F or lower) the koi herpesvirus will become dormant and losses should be reduced. As the water temperature rises the virus becomes active again, but does not tolerate temperatures over 27°C (80°F). However, this does not mean that koi already infected with KHV are free from the disease; simply that they are being kept at temperature levels where the disease will not develop.

Temperatures of between 18°C (64°F) and 25°C (77°F) are the preferred range for the virus that causes KHV to be expressed. For most koi keepers these are summer temperatures and this is the time to be most vigilant, especially after introducing new koi. Bear in mind that it is probably not temperature alone that triggers an outbreak of KHV; usually there is an underlying stress factor, such as low dissolved oxygen concentration or poor water quality, that results in an outbreak of this disease. As KHV will only cause fatalities in a limited band of temperatures, and because the virus can lie dormant within a fish for a period of time, there is no way of being 100 per cent certain that a koi is free from KHV.

So what is the answer? Once KHV is positively identified, should you take drastic action and kill all your stock and disinfect everything? This may seem like the logical action and to some people it will be the right thing to do. However, when you re-stock there is nothing to prevent you from inadvertently introducing more koi with the KHV virus. So an alternative approach is to raise the pond temperature to around 25°C (77°F) and see if the virus breaks out. If it does, let it run its course, removing dead koi as they succumb. Generally not all the koi will die and those which survive may have developed their own immunity against the virus. However, these koi will probably carry the virus, so any new koi introduced during the active temperature range may develop KHV, while any subsequent stress will reduce your koi's immunity, thus risking a "reawakening" of the virus. The other problem with this approach is that you will limit movements from your pond as no fish can be moved because of the danger of spreading KHV. Other than keeping the pond very cold – around 7°C/mid-40s°F or lower – or very warm – 27°C (80°F) or higher – there is no real answer for dealing with KHV at this present time.

In fact, although it may seem drastic, the best possible course of action to take once KHV is positively identified is probably to sacrifice your stock and start again, keeping your fingers crossed that none of the new purchases has KHV. As more people learn about KHV, it is to be hoped that koi breeders, dealers and retailers will take proper steps to test stocks for the virus. When quarantining stocks, many dealers will actually raise the temperature to the range in which KHV outbreaks occur, in order to rule out the presence of this disease.

SVC – Spring Viraemia of Carp

The koi keeper is unlikely to encounter this notifiable disease, since SVC is widespread amongst wild carp in parts of Europe and the Far East and has very occasionally been identified in goldfish reared in polyculture. The only way koi can become infected with the virus is through the introduction of carp expressing the virus – a good reason for not mixing wild carp and koi, especially if you live in a country where SVC is present. In fact it is illegal in most instances to take and move wild stocks without first discussing your intentions with the appropriate authority. It is also wise to avoid importing koi from a region or country that is known to have SVC. That is why strict licensing is in place in countries where SVC is not currently a problem.

SVC is a virus (Rhabdovirus carpio) and can only be accurately identified by sacrificing a koi and sending it away for testing by a laboratory with the facilities to check for this condition.

Generally, an outbreak of SVC is accompanied by haemorrhaging on the body surface, as the virus tends to make the blood vessels become leaky, allowing blood to leach out into the muscle and tissues. Dropsy is a common feature of SVC and infected fish tend to hang in the water near inlets or where the aeration is greatest. Often, the infected fish produce what are known as 'trailing faeces' – in effect, passing mucus. The signs of infection with SVC are also similar to those of other less serious diseases, so always seek professional advice in the event of large numbers of koi becoming sick.

If you are unfortunate enough to encounter this disease in your fish, it is vital that the correct authorities are notified. In most cases the laboratory carrying out the testing will do this as a matter of course. Although the subsequent culling may be hard to bear, try to remember that it is not only in your best interests but also those of the whole koi industry.

Below: One of the many symptoms of KHV is the occurrence of severe gill damage and disintegration of the gill filaments, as pictured here.

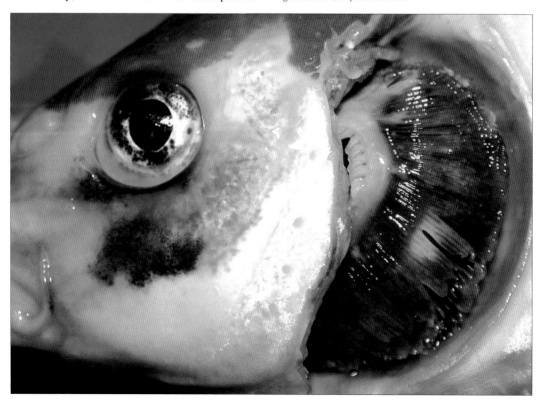

LEECHES

Leeches can be a problem in a koi pond, especially in a planted one. They normally find their way into a pond on plants taken from existing or natural ponds, infected wild fish, or, very occasionally, on amphibians or wild fowl. There are many species of leech, but the most common fish-parasitic leech is Piscicola geometra, which can grow up to 7.5cm (3in).

Leeches attach themselves to a host koi with their sucking mouthparts. These are used to draw blood as a food source from the host koi. In the process leeches can transmit other infections to the host koi. Leeches can also live without a host for a considerable time and will leave the host

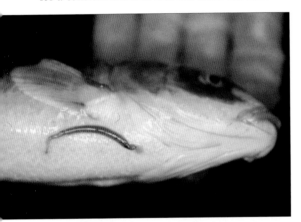

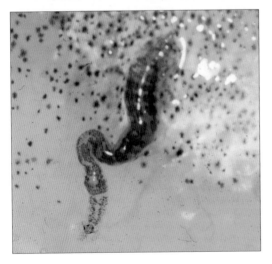

Above top: A leech can be seen attached to the underside of a koi. This should be carefully removed ensuring that the attachment organs are not left embedded in the fish.
Above: Leeches should not pose a particular threat unless they are found physically attached to your koi.

koi to reproduce. They lay numerous eggs which are wrapped in a cocoon which is normally dark brown or grey colour. It may be found attached to plants, rocks, or even the filter media if the leeches manage to make their way into the filter system. Once these eggs hatch, the new leeches find a fresh host and feed on this for a time before leaving the host koi to digest their meal.

Identification
Leeches may be present in a pond for some time and go unnoticed if not actually observed on a host koi. Most species of aquatic leeches are harmless to fish and do not attach to or feed on koi. However, if one is seen on a koi, it is unmistakable – large in size and worm-like in appearance. Leeches attach themselves anywhere on the body of the koi, and a tell-tale sign of a leech infection is the presence of red lesions on the skin, which are caused by the leech attaching itself and drawing blood. If a leech infestation is suspected, it is important to check the underside of the koi; leeches will attach themselves here quite often. Areas such as the face, mouth and the ball joints of pectoral fins are also highly prone to attack.

Prevention
It is important to inspect all new plants for leeches and their eggs before they are introduced to a pond. Unfortunately, however well each new plant is inspected, it is virtually impossible to ensure that all possible hiding places on the plant are checked thoroughly. The fish leech is entirely parasitic on fish; other species of leech that feed on amphibians or wild fowl are rarely found in a koi pond. The only way the fish leech can be introduced is through infected fish, which is rare within the ornamental industry, or on plants that have been cultivated in the wild. Most plants for the ornamental industry are propagated in nurseries and never come into contact with fish. There are some leeches that infect frogs or toads and may also feed on fish. However, it is unusual to find amphibians in koi ponds, which focus on the beauty of the fish and lack aquatic vegetation.

Treatment
It is very hard to eradicate a leech infestation entirely once it has taken hold. In recent years, new products based on mebendazole have been formulated that will help control leech infections.

The life cycle of the fish leech (Piscicola geometra)

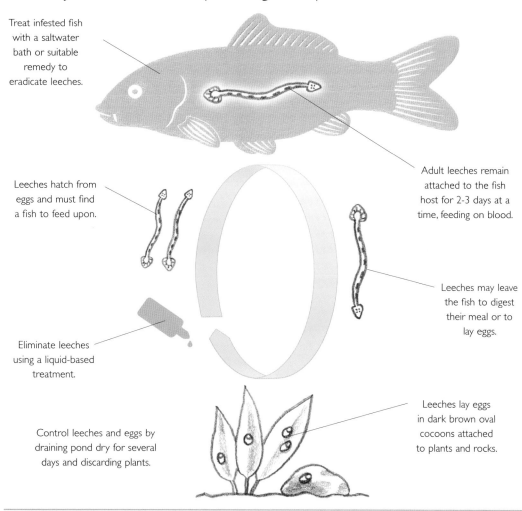

Treat infested fish with a saltwater bath or suitable remedy to eradicate leeches.

Leeches hatch from eggs and must find a fish to feed upon.

Eliminate leeches using a liquid-based treatment.

Control leeches and eggs by draining pond dry for several days and discarding plants.

Adult leeches remain attached to the fish host for 2-3 days at a time, feeding on blood.

Leeches may leave the fish to digest their meal or to lay eggs.

Leeches lay eggs in dark brown oval cocoons attached to plants and rocks.

Normally the treatment has to be repeated many times to ensure complete eradication, as the eggs are not always killed and further treatments are required to kill the leeches as they hatch. In the past, dip treatments were commonly used for controlling leeches but really these are ineffective, as the koi become re-infected when placed back into the koi pond. However, these are not a long-term solution to the problem of leeches as they will only kill leeches on the treated koi and not those still in the pond, or ones hatching from eggs. The only sure way of completely eradicating a leech problem is to drain the pond and all filter systems, and allow the site to completely dry out. This kills all eggs and leeches. While this is taking place, every fish removed from the pond must be inspected and any leeches removed. Leeches have small jaws that they use for biting and these are located quite deep in the pharynx, which is everted to feed. Make sure that nothing is left behind if the leeches are physically removed. Any attachment organs that do remain should be plucked off using tweezers. As the leeches are taken off, it is worth treating the area with propolis to prevent secondary infection. Once returned to the pond, a careful eye must be kept on these areas for secondary infections. If they appear, a further topical treatment should be applied, and advice sought from your vet or local koi health specialist.

MYXOZOA (NODULAR DISEASE)

Nodular disease occurs when koi become infected with myxozoan parasites. This disease can affect koi externally on the body, fins and gills, or internally when organs and muscles are affected. Gill and internal infections are the most harmful, as these go undetected, and often, secondary infections mask the underlying myxozoan infection. There does not seem to be host specificity amongst many species of myxozoa. They have complex life cycles that are not fully understood, but involve fish and, usually, oligochaete worms or a group of organisms known as bryozoa in order to complete their life cycle. In most instances, koi would have become infected with myxozoa while in mud ponds before being exported. Koi ponds with biological filtration are unlikely to harbour oligochaete worms, although bryozoa may be found in re-circulating systems.

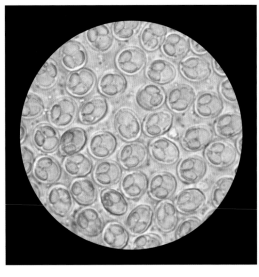

Above: *Myxobolus* seen under a microscope at a high-power setting. The two round capsules clearly visible inside each parasite are an identifying feature.

Identification

As the name nodular disease suggests, lumps appear either externally or internally. These lumps – or cysts as they are correctly termed – may be up to 5mm (0.2in) in diameter and have a white or yellow appearance. They are not uniform in shape; some are circular while others may be elongate or irregularly shaped. Release of spores may be either directly into the environment when cysts on the body surface rupture, release with excretory products or through decay of

Above: This koi is showing swelling around the gills which is exaggerated on one side.

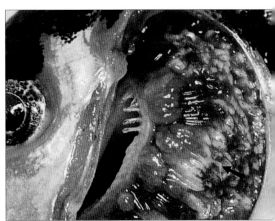

Above: A koi with an advanced gill infection caused by the *Myxobolus* parasite is shown.

Above: Large numbers of *Myxobolus* cysts can form on the gill tissue of infected koi, affecting breathing and causing the gill covers to flare, as seen in this koi.

an infected fish. The spores usually infect an oligochaete worm or the bryozoa and when koi consume these infected invertebrates, the life cycle continues. Myxozoan infections are not contagious from fish to fish and must involve the intermediate host.

The two most harmful instances of myxobolus occur when the gills are infected. Infection of the gills is characterized by white lumps, or cysts, which are the encysted spores. These lumps can be very numerous and large in size. Generally, the cysts are very fragile and little effort is required to break them, releasing the spores. Possibly, friction from the water passing over the gill tissue is sufficient to rupture them. When extensive numbers of gill lamellae bear these cysts, breathing may be affected and the koi may

show signs of respiratory distress, as it is unable to absorb sufficient oxygen. Clinical infections with myxozoa do cause mortalities, but often the underlying cause is difficult to determine, as secondary infections of fungus and bacteria are commonly associated with these parasites.

Prevention

Few koi ponds with re-circulating, biological filters will provide the environment for oligochaete worms to thrive. Some poorly maintained systems may allow the bryozoan invertebrates to produce colonies that can act as intermediate hosts for the myxozoa. In most instances, a koi infected with myxozoa was already infected when purchased. However, these parasites are impossible to see unless the gills bear the characteristic white cysts. Often, koi may have myxozoan infections but there are no outward signs of disease and, even after the cysts have ruptured, the koi may well live a long and healthy life. Obviously, it is important to ensure that pond hygiene is excellent and that the water is free of organic detritus, which encourages the growth of bryozoan colonies. It is also sensible to avoid using live food, as this could be a secondary host for spores released by already infected fish. If you do want to feed live food, it may be best to opt for a frozen type, as these are generally treated and sterilized to destroy pathogens.

Treatment

There is no effective treatment for koi infected with myxozoa, as a large part of the life cycle takes place within the cells of the infected koi and in most instances the parasite is only recognized when the white cysts are evident. On the whole, infection with myxozoa tends to be a chronic disease that usually goes unrecognized and often has little effect on the koi, which appears healthy. Occasionally, outbreaks of myxozoan parasites do indeed result in mortalities, but these events are often misdiagnosed as bacterial or due to a secondary infection with another parasite. The best treatment is prevention by maintaining appropriate husbandry and water quality, regularly cleaning the mechanical filtration and keeping it free from organic detritus. The key to inhibiting further infection with myxozoa is to break the life cycle by preventing the intermediate hosts – in koi ponds these are bryozoa – from colonizing the filtration system.

SKIN GROWTHS AND CARP POX

Lumps and bumps on the skin of koi may be referred to as 'papilloma', although the histological origins are unidentified. Carp, or fish, pox is due to a viral infection. Young koi less than one year old tend not be affected by skin growths and older koi sometimes seem to grow out of it. This is a relatively harmless disease and does not pose a major threat to your koi. Carp pox has a very low level of infection and so one fish in a pond may show symptoms while the others remain healthy.

Identification

Carp pox is easily identified by the hard, white, waxy lumps which appear mostly on the fins of the fish and occasionally on the head. On Doitsu or leather koi, carp pox may also be found on the body as it tends to be found on areas lacking scales. Carp pox is often described by professionals and hobbyists alike as if a white candle had been lit and the hot wax allowed to drop onto the koi. Skin growths are similar in appearance to carp pox with large areas appearing on the fins, especially around the hard leading ray. Elsewhere they tend to be smaller with an average size of around 5mm (0.2in), and may be present all over the koi in quite large numbers in some cases. Like carp pox, they may have a white waxy appearance; however, you may also notice tumour-like growths which look red or pink.

Prevention

The reasons why koi may develop various tumour-like growths on the body are unknown, which precludes prevention. Carp pox is generally associated with cold water conditions. If your koi suffer an outbreak of carp pox, it should not be any real cause for alarm as this viral infection is generally regarded as harmless. The worst effect is the unsightly appearance that your koi take on while the virus is active.

Treatment

There is no cure for carp pox; however, an increase in water temperature may cause the virus to subside and the white waxy lumps to disappear, as the increased water temperature enables the koi to mount an effective immune response against the virus. An increase of around 10°C may be sufficient, however care should be taken to implement this increase over a period of days and even weeks with an increase of 1°C being made every one to two days. Rapid temperature increase should be avoided as this will be very stressful for the fish and could result in other diseases occurring, especially parasite infections such as whitespot. The symptoms of carp pox may well disappear with this increase in temperature, but this does not mean that the koi in question is free from the disease. White waxy lumps may suddenly appear again in the

Above: This fish has carp pox on the head and face, which is one of the most common areas for it to be found. Despite being unsightly, it will seldom prove a serious health threat to your koi.

Above: A close-up of a pectoral fin showing a small area of carp pox. Fins are areas which are prone to carp pox.

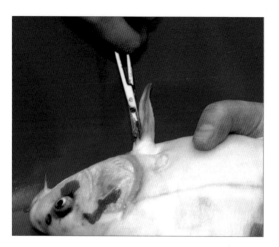

Above: A skin growth is being removed from the pectoral fin of this koi, a task which should be performed by a vet.

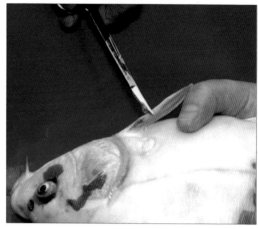

Above: The last sections of the growth are excised using sterilized scissors.

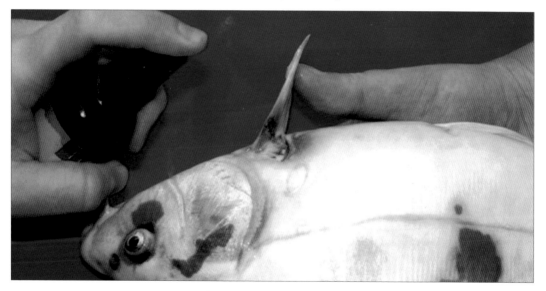

Above: After surgery, topical treatment is applied.

future, generally during a period of lower water temperature.

Growths or tumours on the skin may be difficult to treat, as the underlying cause is hard to identify. Unlike carp pox, raising the temperature will have little effect, and if the tumour is malignant (i.e. cancerous in its origin), heating the water and increasing the koi's metabolism could cause the growth to proliferate much more rapidly. As there are various origins of skin growths, they can appear quite different. They may look quite flat, but on closer inspection

the skin is quite different in appearance from that surrounding it. They may have a warty appearance and very occasionally quite significant tumour-like growths are initiated following an outbreak of carp pox. These may need to be removed surgically, a job which should be done by a vet or your local koi health specialist. The areas from where these tumours are removed should be topically treated with malachite green and propolis, for example. In extreme cases when a large number are removed, secondary infection may occur which will require additional treatment. Advice should be sought from your local koi health specialist in this case.

POP-EYE (EXOPHTHALMIA)

This is a condition which affects either one or both eyes, and causes them to stand out from the body. If just one eye is pronounced, the chances of a serious problem and losses resulting are much less than if both eyes are affected as this may suggest a more serious (and possibly internal) problem. Pop-eye can be caused by a number of factors including viral infections, bacterial infections, poor water quality, nutritional deficiency, internal problems and even physical damage, such as badly netting the koi, or as a result of spawning.

Identification

Pop-eye is not usually a highly infectious condition and so only one koi may show symptoms at any one time; however, the underlying cause will determine if the condition will prove infectious or not. The tell-tale signs of pop-eye are unmistakable – the eyes stand out from the body, and in extreme cases this can be by over 10mm (0.4in). Remember that a koi's eyes normally protrude slightly from the body, so before assuming that a fish has pop-eye compare it with a known healthy koi. If it is pop-eye, it is vital to establish what the cause of the problem is. Stress is the most common trigger and poor water quality is the main culprit. This should be the first avenue to investigate. If water quality proves fine, think about any changes that have occurred; even events like the introduction of one or two new koi can be enough to trigger pop-eye in some fish.

You may also find that a particular koi develops pop-eye at certain times of the year every year, and this may simply coincide with seasonal changes in water temperature. If you also notice that areas of scales are lifting, or if the koi has gone off its food, is starting to waste or there seems to be swelling, it could be that a bacterial or viral infection is causing the problem. It could also be linked to the onset of dropsy. Internal problems are another trigger and these can be anything from a build up of fluid, to organ failure or even a tumour. As these causes are internal, it is often difficult to identify them precisely. As long as they do not result in the onset of a more serious bacterial infection or even dropsy, the condition may still clear up of its own accord.

Prevention

Simple good husbandry is a must – if excellent water conditions are maintained at all times, and other stress factors are avoided, the likelihood of pop-eye developing are dramatically reduced. Using a pond heating system can also help in reducing stress, as a constant temperature can be maintained throughout the year. Offering a good, well-balanced koi food will help prevent this condition. Once opened, store the food in an airtight container to prevent deterioration of the vitamin content, which can lead to conditions such as pop-eye. When buying new koi, avoid any showing signs of pop-eye.

Treatment

As pop-eye is generally not contagious, it is not important to isolate the affected koi. In fact the koi generally improve faster if kept in their normal environment as this will generally have better filtration and be more stable than most quarantine or isolation facilities. Try to get at the underlying cause. Often a simple water test will

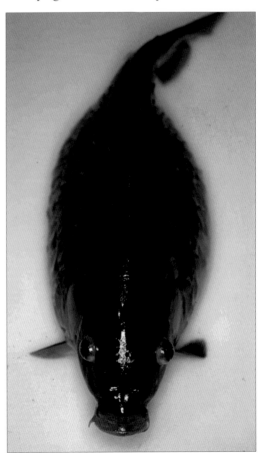

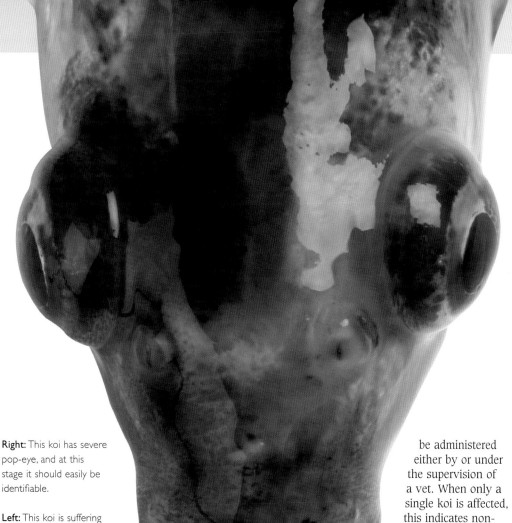

Right: This koi has severe pop-eye, and at this stage it should easily be identifiable.

Left: This koi is suffering from an advanced stage of dropsy. The eyes are clearly protruding from the profile of the head.

show a high level of ammonia, and once this has been remedied via small water changes with conditioned, dechlorinated tapwater allied to a reduction in food levels until optimum water conditions are re-established, the condition will normally cure itself.

Once this, and other possible causes such as physical injury, have been ruled out, it may be necessary to treat with anti-bacterials as a bacterial problem may be behind the outbreak. If bactericides are required, seek advice from a qualified professional as to what should be used, and at what dose, and how it should be administered. In most cases where the condition is due to physical injury, this will heal naturally, but if you suspect an infection, then it is important to seek professional help. The koi may need an antibiotic drug, which should only

be administered either by or under the supervision of a vet. When only a single koi is affected, this indicates non-infectious disease. Pop-eye caused by bacterial or viral disease spreads rapidly to affect most of the koi in the pond. You may wish to treat the pond with an anti-bacterial medication such as Acriflavine at the dose recommended on the bottle (as this will depend on the strength of the solution mix), or potassium permanganate at the dose of 1.5g per 1000 litres (220 gallons). Use with caution where the pH is alkaline. Chloramine T at the dose of 1g per 1000 litres (220 gallons) is another treatment, but avoid using it in soft water areas. Alternatively, use an off-the-shelf product intended for bacterial infections. If both eyes are protruding and you notice scale lifting and swelling, the koi may have dropsy; follow the recommended treatments for dropsy. Another option is to give the fish a salt bath of 100g (3.5oz) of salt per 4.5 litres (1 gallon) of water for 10 minutes. This can be done once a day for three consecutive days.

SWIMBLADDER DISORDERS AND AIR GULPING

A fish's swimbladder is the organ used to maintain neutral buoyancy in the water, and a koi with symptoms of a swimbladder disorder will have difficulty in swimming normally. Swimbladder problems can occur simply of their own accord or they can be triggered by another health problem, such as a tumour, internal fluid build-up, or even bacterial infection. In such cases once this problem is cured, the chances are that the swimbladder problem will also disappear as long as no internal damage has occurred to the organ itself. There are many other reasons why koi may develop disorders of the swimbladder and in most instances these are very difficult to identify without resorting to veterinary intervention and radiology. Even though an X-ray may be helpful in identifying the reason why the swimbladder is diseased, there may not be a cure. Contrary to popular opinion, temperature has little effect on the functioning of the swimbladder.

Identification

Air gulping occurs when a koi takes in too much air to the extent that it becomes bloated. As a consequence it may start to show the same symptoms of a swimbladder disorder. This condition, however, normally corrects itself after a short period of time.

Typically affected koi either spend long periods of time on the bottom of the pond or seem to float permanently at the water surface. Koi belong to a group of fish known as 'physostomes', which retain a special duct that connects the swimbladder to the oesophagus. Healthy koi can 'top up' the swimbladder by gulping air. Shipment and transportation for a few hours causes changes in dissolved gas pressure in the transport water and can lead to koi breaking the water surface to gulp air periodically to adjust the gas content of the swimbladder when introduced to the pond. Normally, the gas content of the swimbladder differs from air, in that the buoyancy organ contains a greater volume of gaseous oxygen. Normally, blood vessels in contact with the swimbladder epithelium put gas into the buoyancy organ and gas can be released either by burping or via the intestine. The first sign of swimbladder disease tends to be associated with the koi swimming to the water surface and gulping air. Affected koi have difficulty swimming because they are no longer neutrally buoyant but heavier than water and therefore swim through the water with a characteristic rowing action. As soon as they stop swimming, they tend to sink to the bottom of the pond. If the koi is unable to reach the surface to take food, or if it spends large amounts of time on the pond bottom and develops pressure sores that often become secondarily infected by bacteria, it may be necessary to consider euthanasing it.

Prevention and Treatment

Because it is difficult to identify the reason why koi develop swimbladder disorders, successful treatment of the condition is unusual. Taking X-rays and calling for veterinary intervention to aspirate air or fluid from the swimbladder just relieve the signs of the disease, but do not treat the underlying cause. Female carp – and therefore koi – are very prone to developing fibrous cysts in the ovaries. Over time, these appear to worsen and often cause pressure on either the blood vessels supplying the swimbladder or the duct that connects the

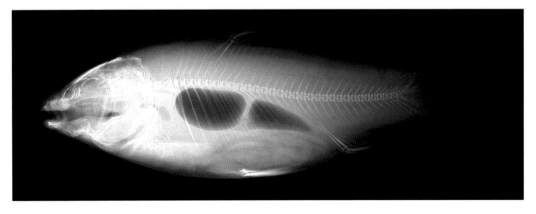

Above: A koi with swimbladder problems may not just sink or float within the pond; it may also have trouble maintaining an upright position in the water, as shown here. For a conclusive confirmation that the swimbladder is the cause of this, a vet may have to perform an X-ray.

buoyancy organ to the oesophagus, leading to the loss of buoyancy.

Another measure to help the koi is to transfer it to a shallow pond, such as a quarantine facility. A shallow pond reduces the weight of the water pressure on the affected koi but it is important

to realize that the fish is unlikely to recover and may well need to spend its remaining days in the quarantine pond. Salt baths are another option – a dose rate of 22g per litre (3.5 oz per gallon) should be used for ten minutes. This can be done up to three times, with a bath being given on three successive days. Should bathing prove to be impractical, salt can be added to the pond at a dose rate of 3g per litre (0.5oz per gallon). However, be sure to test the salt levels and ideally reduce them to zero before using other medications, such as formalin. Formalin and salt work in the same way by stripping parasites and mucus from the koi's skin, and if you use both at the same time the fish may be burnt. If the affected koi is of particular value, it may be worth locating a specialist fish vet as he may try to reinflate or deflate the swimbladder. However, this procedure may have only limited success.

With regards to air gulping, the best approach is to limit situations in which air is taken in, which happens most commonly during feeding. Nowadays a feed mix comprising 50 per cent floating and 50 per cent sinking food is recommended and this combination should help to reduce the occurrence of this problem, as no air will be taken from the surface when the fish consume a sinking food. Indeed you might want to use only sinking food, but take care then to ensure that the food is not extracted from the pond by bottom drains or pumps before the koi get a chance to eat it.

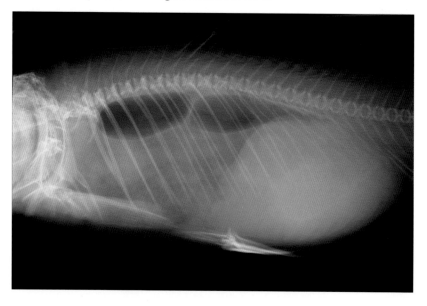

Left: This X-ray shows the swimbladder of a healthy koi – a two-chambered organ located at the top of the abdominal cavity.

Right: The swimbladder of this koi has become compressed by a tumour, which is seen as a shadow above the pelvic fins in this X-ray.

TAPEWORM INFESTATIONS

There are only a few species of tapeworm, but the main species that affects koi is Bothriocephalus acheilognathi. When this tapeworm infects young koi it is very serious, as B. acheilognathi is usually present in large numbers and completely occludes the intestine, depriving the fish of nutrients. Signs of infection with this tapeworm are extreme: either the koi appears to have a very bloated abdomen or it is feeding voraciously but is emaciated. In very heavy infections, the proglottids (segments) may be seen protruding from the vent. Obviously, you should avoid buying any koi showing these signs of an advanced infestation.

Life Cycle

Bothriocephalus are parasitic worms and in their adult stages they attain lengths of 15-23cm (6-9in) and have a body width of 3mm (0.1in) within the intestine of a koi. They have a white ribbon-like appearance, but are generally only seen when a fish undergoes a post-mortem. Before they reach this size they must complete a complex life cycle. Firstly a fish already infected with a tapeworm excretes waste containing eggs of the internal worm into the water, and after a time they turn into free-swimming larvae. These newly hatched larvae must then find a host in which to develop further. This host is often

The life cycle of a tapeworm (Bothriocephalus)

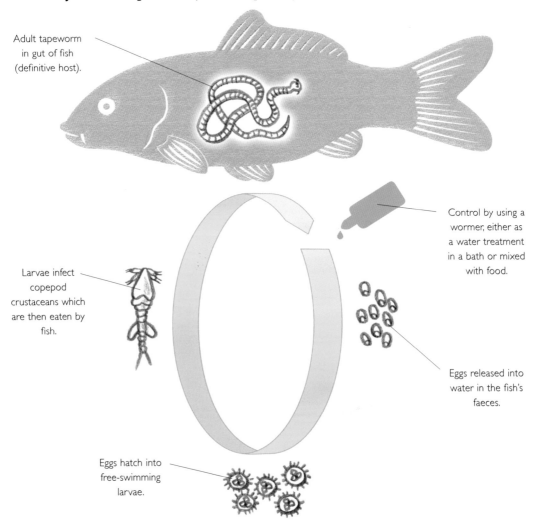

Adult tapeworm in gut of fish (definitive host).

Control by using a wormer, either as a water treatment in a bath or mixed with food.

Larvae infect copepod crustaceans which are then eaten by fish.

Eggs released into water in the fish's faeces.

Eggs hatch into free-swimming larvae.

Right: The typical koi hobbyist will be very unlikely to see a tapeworm. Generally they are only observed when a post-mortem is performed.

Above: Tapeworms in koi can reach up to 23cm (9in) in length inside the intestine of the fish.

a copepod which may simply eat the larvae, which triggers the next stage of development. A copepod is a free-living crustacean (and hence is related to anchor worms and freshwater fleas), which may be naturally present in any body of water, but in a pond they are generally introduced on plants or live food. If a koi eats a copepod infected with B acheilognathi, the fish will become infected with the tapeworm.

Prevention

The best preventative measure is to avoid the introduction of any koi which already have an infestation of Bothriocephalus sp. But as there are seldom external signs of infection, in reality this can prove very difficult. The next step to take is to ensure that the copepod host is not present. Fortunately, copepods are rare in the koi pond as the lack of food in the form of plankton is absent, although possibly the copepodid stage of anchor worm could act as an intermediate host. You can avoid adding plants to your pond, and steer clear of live food as both may harbour copepods. If you do favour plants and live food, make sure that you disinfect them first. It can prove difficult to detect infection with B. acheilognathi, but fortunately, because the intermediate copepod hosts are rarely found in the koi pond, the tapeworm will not spread

to other fish in the pond. The tapeworm does have a finite life span and will eventually die, whereupon the tough integument (external covering) that protects it during life breaks down and the koi can take its revenge by digesting it.

Treatment

As we have seen, in a well-maintained pond the intermediate hosts needed to complete the life cycle of tapeworms are not always present, so treatment is not needed. Overall these worms are a small threat to the average pond keeper and as long as a good, clean and well-maintained system is kept they are unlikely to be encountered. They are more of a problem for the fish farmer and breeder. Should you feel that treatment is required, you may wish to use a proprietary worming treatment either as a water treatment at the rate recommended by the manufacturer, or mixed with the food if the medication is unpalatable to the koi. The drug praziquantal is most commonly used to eradicate tapeworms in fish. In the past, it was used in an injectable form, which was not especially effective in controlling worm infections. Recently, a soluble form of this drug has been introduced. It is available as a pond treatment and has proved to be very effective. Soluble praziquantal is only available through a veterinary practice.

TRICHODINA

Trichodina is a commensal rather than a parasite; it feeds on bacteria and organic detritus and the presence of koi are unnecessary for it to complete its life cycle. Trichodina is found more profusely in poorly maintained systems where mulm and sediment have been allowed to build up in the pond and filter.

Identification

Trichodina is only visible under a microscope, as its average size is just 0.07mm (0.003in). It looks like a small circle fringed with a number of denticles inside a larger circle and it may be spinning and moving at quite high speed. Trichodina are found on the body surface and gills of the koi. The calcified denticles merely support the soft body structure. It moves around the surface of the koi by means of hair-like cilia that are flexed in a wave-like motion. The mucus of

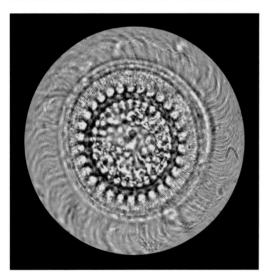

Above: Seen under a microscope, *Trichodina* looks like a small circle with denticles attached and a fringe of cilia.

the koi traps bacteria and other organic detritus, which encourages Trichodina to infect the fish. The presence of Trichodina irritates the koi, which produces more mucus that traps more detritus and thus the cycle continues. Ultimately, the mucus cells of the skin become exhausted and the koi's body becomes dry and prone to bacterial and fungal infection.

Trichodina will readily swim from koi to koi, and, if conditions allow, will multiply quickly by a process of division. Mature infected koi may not show any obvious signs of disease despite having a heavy infestation. Koi under one year of age are more susceptible to severe Trichodina infestation and they may exhibit numerous symptoms. The initial indication that something is wrong is generally flicking and rubbing as the fish tries to relieve the irritation caused by these parasites. This behaviour can lead to physical damage which is vulnerable to secondary infection from bacteria and fungus. Younger koi (under one year) will show symptoms very quickly while koi aged two years and more may not show any adverse signs until the infestation levels become much higher.

Once any koi show signs of irritation through flicking or rubbing, it is sensible to take a sample of mucus on a slide and examine it under the microscope to obtain a positive identification that Trichodina is responsible. You may also notice that the koi develops a whiteish appearance, especially around the head area. In the early stages of infection, special cells in the skin produce extra mucus to try to reduce the irritation, but this actually traps more food, which in turn encourages the Trichodina to thrive. Left untreated, the skin cells become exhausted and the koi becomes progressively weaker, usually succumbing to secondary infections. Trichodina can also infect the gills and then you may well see your koi seeking the oxygen-rich water around waterfalls or airstones. You may occasionally also see the koi shake its head rapidly as if it were trying to remove

Below and above right: A koi with a *Trichodina* infection – the skin tissue around the eye has been eaten away.

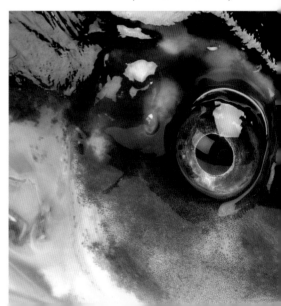

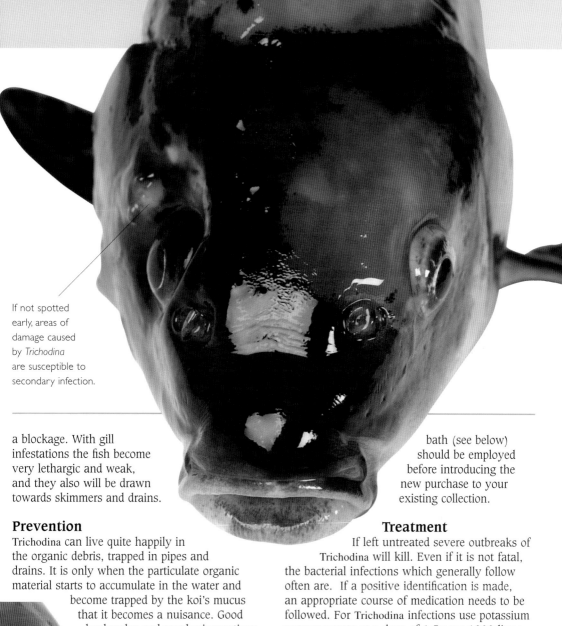

If not spotted early, areas of damage caused by *Trichodina* are susceptible to secondary infection.

a blockage. With gill infestations the fish become very lethargic and weak, and they also will be drawn towards skimmers and drains.

Prevention

Trichodina can live quite happily in the organic debris, trapped in pipes and drains. It is only when the particulate organic material starts to accumulate in the water and become trapped by the koi's mucus that it becomes a nuisance. Good husbandry and regular inspections of the filters also ensure that only limited amounts of debris are allowed to collect, as debris build-up can contribute to heavy infestations. Since Trichodina may be introduced with new fish, avoid purchasing any koi showing the symptoms of Trichodina. If in doubt, a period of quarantine or a precautionary

bath (see below) should be employed before introducing the new purchase to your existing collection.

Treatment

If left untreated severe outbreaks of Trichodina will kill. Even if it is not fatal, the bacterial infections which generally follow often are. If a positive identification is made, an appropriate course of medication needs to be followed. For Trichodina infections use potassium permanganate at a dose of 1.5g per 1000 litres (220 gallons), but use with caution where the water has a high pH. This can be repeated if required at five to seven day intervals for three weeks maximum.

An alternative to potassium permanganate is to give a two per cent salt bath – 20kg of salt per 1000 litres (44lb per 220 gallons) for ten to twenty minutes. It is also important that any secondary infections are topically treated with propolis plus a suitable bactericide. If severe ulceration starts to appear, further treatment may be required and this should be discussed with your vet or koi health specialist.

WHITESPOT

Whitespot is caused by the protozoan parasite *Ichthyophthirius multifiliis*. Most koi keepers encounter this disease and it is easy to deal with if identified early. Infection is generally stress related; the parasite takes advantage of the koi while the immune system is suppressed. Although whitespot is considered to infect koi in the spring, the non-specific immune system is still effective and will generally repel invading whitespot in healthy koi. The immature whitespot burrows beneath the koi's epidermis, consuming the cellular debris that it produces. The immature whitespot remains within the skin, where it continues to feed and grow to 1mm (0.04in) or more. Once sufficient lipid and protein have been absorbed by the parasite, it appears to release proteins that disrupt the skin, allowing the parasite to escape.

Above: Excess mucus is a response to the irritation caused by the whitespot parasite.

Whitespot reproduces by division, and within this cyst many hundreds of new whitespot parasites are produced. At this early stage they are highly infective. These new whitespot or "swarmers" then leave the cyst and become free-swimming as they try to find a new host to infect. If no host is found within 48 hours at 20°C (68°F) they will die. If a new host is found, the whitespot will cause further infection, or re-infection, by breaking through the top layer of skin,. They will remain in situ for up to 20 days (possibly more), before the cycle starts all over again. This life cycle is governed by temperature – the complete process takes longer at lower temperatures and in winter months, but the whitespot parasite is able to become dormant for long periods of time, possibly years. By contrast, at higher temperatures the whole life cycle is speeded up – it can take place in as little as seven days at temperatures of 21°C (70°F).

Identification
Very early stages of whitespot may not produce any visible external signs other than changes in your koi's behaviour. You may notice that a fish starts to clamp fins, becomes lethargic and occasionally rubs against the bottom and walls of the pond. If you notice these symptoms, take

a skin scrape to allow accurate diagnosis. In later stages of whitespot disease, a few white spots may appear on the koi, and over time these increase in number so that it looks as if salt has been sprinkled over the koi's body. As more spots appear, koi may start to lose their appetites and rub vigorously against objects in the pond to try to relieve the irritation caused by the parasites. They may also start to hang in the water and spend more time near areas of higher oxygen, such as airstones. At an advanced stage of the disease, the whole body of the koi may take on a whiteish appearance as it produces large amounts of mucus to try to relieve the irritation. The intensity of the flicking may increase, causing numerous physical lesions to appear, and secondary bacterial and fungus infections can set in. At this advanced stage the mature parasites start to leave the koi to reproduce, and areas will be exposed on the body which will be a target for secondary infection. If the disease has reached this stage, you will probably not only be dealing with whitespot, as other parasites and pathogens will take advantage of the situation and start to infect the koi. It is sometimes these secondary infections that kill the koi, if no suitable treatment is given.

Prevention
Good husbandry and water management will help to keep the occurrence of this disease to a minimum. Whitespot is also triggered by water temperature change and is a common problem after new koi are added to a pond, as the water temperature is usually different and they will be stressed from transportation. Changes in water temperature are a major contributor to the occurrence of whitespot so the installation of a heating system for your pond will help in maintaining a stable water temperature. Whitespot may also be introduced on plants, so always disinfect these before adding them to the pond. This can be done by dipping the roots of plants and immersing live food in a potassium permanganate bath of 0.8g per 4.5 litres (1 gallon) for five minutes. However, this may have a

detrimental effect on more delicate plant species, so it is better to avoid adding plants and live food if possible. Alternatively keep them in a fish-free environment for a week at a temperature of 21°C (70°F) or above as this will cause any cysts to develop into swarmers which will soon die in the absence of fish to infect.

Treatment

Unfortunately when whitespot has reached its mature stage at which it is visible, it is relatively immune to treatment as it is embedded between the top two layers of skin. So when treating whitespot you must aim to kill the free-swimming stages. There are many off-the-shelf medications formulated for treating whitespot and these can be used with good results. Alternatively malachite green can be used – the dose rate will depend upon the concentration of the mix. In the past it has been recommended that raising the temperature to 28°C (82°F) would eradicate a whitespot infection. More recent studies suggest that these temperatures will not eradicate the parasite and, more importantly, such high temperatures are close to the maximum tolerance for the koi. Given that there are efficacious medications on the market for controlling whitespot, it is better to use these than to subject koi to potentially dangerous temperatures. Do remember that secondary bacterial and fungal infections may have occurred and these will need topical treatment with fungicides, bactericides and propolis. Koi that have been infected with the whitespot parasite are immune to re-infection for about two years.

Whitespot life cycle
(Ichthyophthirius multifiliis)

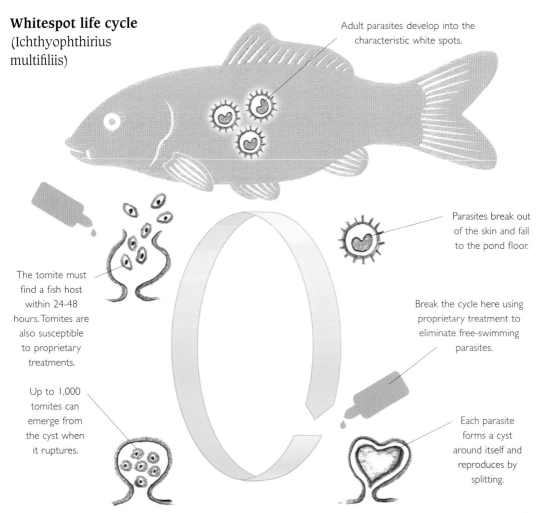

Adult parasites develop into the characteristic white spots.

Parasites break out of the skin and fall to the pond floor.

The tomite must find a fish host within 24-48 hours. Tomites are also susceptible to proprietary treatments.

Break the cycle here using proprietary treatment to eliminate free-swimming parasites.

Up to 1,000 tomites can emerge from the cyst when it ruptures.

Each parasite forms a cyst around itself and reproduces by splitting.

OTHER HEALTH PROBLEMS – AN OVERVIEW

Blood Parasites

These are not often a problem for the general koi keeper, and in fact the chances of them even being identified are slim, because a koi needs to go through an extensive post-mortem for an exact identification to be made. Often there are no outward signs of mild to moderate infections

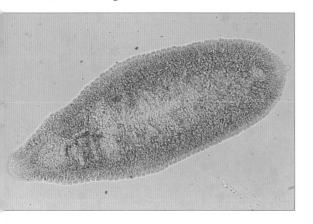

Above: *Sanguinicola* is a blood parasite which is very rarely experienced as a problem by the average koi keeper. Here it can be seen viewed under a microscope.

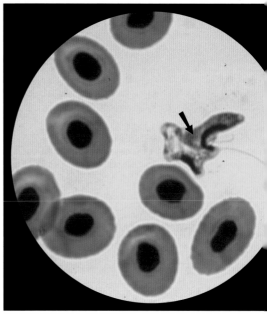

Above: For an exact identification of *Trypanoplasma* to be made a specialist must be contacted. Once again, this is seldom a problem for the average koi hobbyist.

with blood parasites, although heavy infections of trypanosomes may result in anaemia, and blood fluke can cause gill and kidney damage. All these are common symptoms of numerous other diseases which makes positive identification of these parasites even harder. If you want to research blood parasites further, the common culprits are Trypanosoma and Trypanoplasma which can be spread by leeches in a pond. The other blood parasite which may be encountered is the bloodworm Sanguinicola. In its developmental stages this needs an aquatic snail as an intermediate host. So eradicating the presence of all snails will help to prevent this parasite from causing problems. Blood parasites are extremely hard to treat. If a positive identification is made, seek professional advice from either your vet, the laboratory which carried out the tests, or a suitably qualified professional as to the best course of treatment.

Dragonfly Larvae and Frogs

In their larval stage dragonfly live in water, so a pond is an ideal environment for them to grow to maturity. Although rarely a threat to koi, it

is worth checking regularly for the presence of larvae. However, a typical unplanted koi pond is unlikely to be the chosen habitat for dragonfly; they are far more likely to seek out a heavily planted or more natural pool. These creatures can also be introduced on plants, so wash and check these before adding them to the pond.

Very occasionally frogs may also prove a problem in the mating season when they may become over-amorous and try and attach themselves to an unsuspecting koi. Although in most cases the frog will soon realize its mistake and let go, physical damage may occur and this may require subsequent treatment. In very extreme cases it has been said that frogs can attach themselves so tightly to the fish that they actually block the gills and can suffocate the koi – it must be stressed that this is extremely rare. In the unlikely event that you witness a frog attaching itself to a koi, give it a chance to let go of its own accord, and then take steps to remove the frog by hand. It is impossible to prevent wildlife from visiting your pond; all you can do is keep a watchful eye, and perhaps make the pond unattractive to wildlife to discourage it.

Fish Tuberculosis (TB)

This is another disease (caused by mycobacteria) which is very unlikely to affect the average koi keeper. If it ever is identified, great care should be taken when dealing with it, as this disease is a zoonosis, meaning that there is a chance of infection spreading to humans. Signs of a possible infection of fish by mycobacteria include weight loss and the apparent wasting of the infected koi, combined with a drastic loss of appetite. Again, symptoms include many other signs which are also indicative of other more common diseases, and these are more likely in most instances to be the culprits rather than fish TB. These symptoms include areas of reddening on the skin, which may then turn into small ulcerations, pop-eye, excessive periods of hanging in the water, clamped fins, and possibly erratic swimming behaviour. The only way an exact diagnosis can be made is to send away a fish for testing. If you get a positive identification, seek advice from your vet.

It may be necessary to euthanase the infected koi as the prognosis is poor. It is possible that tubercular lesions occur on the skin of the infected koi, and others in the pond may become infected through attempting to 'clean' the infected fish. However, on the whole, the disease does not seem to spread easily to the other stock. Purely as a hygiene measure it makes sense to remove any koi that has died from the pond as quickly as possible. The most likely transmission of tuberculosis results from allowing other fish to feed on the decomposing body of an infected koi. Although many cases may go undetected and undiagnosed, this disease appears to be very uncommon in koi and should not be a major cause of alarm.

Above: Improbable as it may seem, cases have been reported of over-amorous frogs attaching themselves to unwary koi during the mating season. They usually let go without any human intervention being needed.

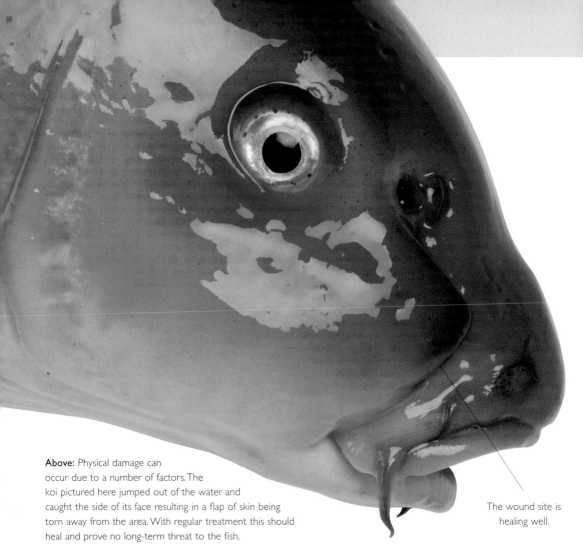

Above: Physical damage can occur due to a number of factors. The koi pictured here jumped out of the water and caught the side of its face resulting in a flap of skin being torn away from the area. With regular treatment this should heal and prove no long-term threat to the fish.

The wound site is healing well.

Physical Damage

This is a common problem for the koi keeper, and can be caused by anything from bumping into sharp rocks, bad netting, spawning activity, and poor transportation to attack by predators, such as herons, magpies or crows. The best step to take to avoid the occurrence of physical damage is to reduce the objects in the pond on which the koi can damage themselves. If predators like herons are a problem, consider netting the pond or installing a suitable deterrent device. If spotted early most physical damage can easily be treated with topical application of malachite green and propolis. However, if unnoticed for any length of time, bacterial or fungal infections may develop and these may need further treatment, possibly including the use of anti-bacterials and fungicides. Obvious signs of physical damage include missing scales, areas of reddening, split fins, and grazes on non-scaled areas, such as around the mouth or all over Doitsu koi.

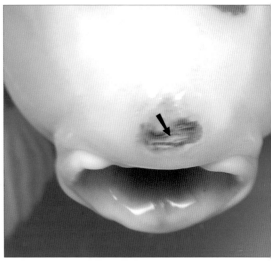

Above: This koi has damage around the mouth which may have happened either in transit (bag rub) or due to feeding around the edge of the pond and rubbing its face.

Sleeping Sickness of Koi

There are probably multiple causes of the condition known as 'sleeping sickness'. For example, koi exposed to high nitrite concentration in the pond water will also lie on their sides and appear lifeless. It is not a clinical disease as such, rather a term that is used to describe a particular behavioural abnormality. This condition usually only affects young koi – koi over two years of age may not show any symptoms at all. Typically the koi appear to be dead, and simply lie on the bottom showing no signs of movement. The cold temperature has the effect of dramatically reducing their metabolism so they may not appear to be breathing as the gill movement is so slight. However, when stimulated the koi will move slowly but only for a short time, before relapsing into their previous condition. This condition can be treated by adding salt to the pond and raising the water temperature to at least 21°C (70°F). If sleeping sickness is suspected, the temperature can be raised rapidly at a rate of 3-4°C every day or two as prolonged exposure to low temperatures dramatically reduces the chances of recovery from this condition. While doing this salt can be added at a dose rate of up to 6kg per 1000 litres (13lb per 220 gallons), but after ten days this will need to be reduced to a lower level through water changes.

Although this condition is sometimes experienced by the hobbyist, it is more of a problem for farmers and breeders who keep vast stocks of young fish. If not treated quickly, losses should be expected.

Sunburn

Koi with areas of white pigmentation are susceptible to sunburn, and this is characterized by a reddening of the white pigment; in extreme cases blistering may occur. The best way to stop this happening is to provide adequate shade over the pond. The whole pond need not be covered but at least an area of it should be. An alternative approach is to build a pergola over

Above: Though they look dead, these fish are just suffering from sleeping sickness. As year-round heating becomes more popular with the serious koi hobbyist, outbreaks of this condition become less likely.

the pond, and use greenhouse shading or shaded polycarbonate to reduce the levels of direct sunlight which can get to the pond. This will have the added benefit of reducing green water and blanketweed growth. If a koi does develop sunburn, avoid the use of harsh medications on the affected area as this may aggravate the situation. The simple application of propolis should be sufficient to clear things up. Just keep an eye on the affected koi to ensure that no secondary infections occur, such as fungus or bacterial infections. Should these arise, take the necessary steps to treat them, i.e. further topical treatment, or in the case of bacterial infection seek expert advice.

TREATMENTS

In this final section the ways in which you can treat your koi will be described and illustrated. This section also explains what medications and treatments are available, what they look like, and how they should be used. Many of the treatments can be obtained from your local koi specialist or vet. Over recent years there have been significant advances in the types of medication available to treat koi suffering from parasite infections, such as fish lice, anchor worm and even tapeworms. The new medications are readily soluble in water and easy to use, and without any long term harmful side effects to either the koi keeper or the environment. As improvements are continually made, they may well prove a good option too.

It is vital that you find a good vet or koi specialist, and form a strong relationship with them to ensure that you can obtain medications and services when needed. If all other avenues of treatment fail, it may be necessary to use antibiotics as a last resort and these will have to be obtained from your local vet. However, before antibiotics are ever used, a full sensitivity test must be obtained by taking a swab. Administering the wrong antibiotic is as bad as giving none at all, and if it is given for a prolonged period it will encourage a resistance to that particular drug to build up. The only exception is when a koi is so ill that the delay in waiting for the swab results is unacceptable. In such a case, intermediate antibiotic treatment is at the discretion of the vet. It must be stressed that antibiotics should not be used without the guidance of your vet or koi health specialist. Do be aware that the treatments mentioned here are all potentially harmful if they fall into the wrong hands, and so all medications and treatments must be kept securely. Make sure that:

• Medicines are kept out of the sight and reach of children and animals.
• All bottles are clearly marked, with the appropriate health and safety information.
• All bottles have a safety lid to prevent children from opening them, should they fall into their hands. **Never** mix up a powder-based medication and, finding that you have some left, decant it into an empty soft-drinks bottle!

• When not in use, the storage area where all medications are housed should be locked.
• After treatment, be sure to tidy away thoroughly making sure that any spillages are cleaned up, and any unused medication and used apparatus, such as syringes, are disposed of in the correct way.

In addition, when using these treatments it is essential that you take all necessary steps to protect yourself and the surrounding area. Many of the treatments will stain anything with which they come into contact and these stains can prove virtually impossible to remove. Any buckets or containers which are used for mixing will be stained, so it is a good idea to keep separate containers which are just for pond use. Furthermore any buckets or watering cans which have contained other chemicals, like weedkiller, should be avoided. You should wear rubber gloves, and ideally an apron or some form of protective clothing. When using some medications, such as those which are powders or give off fumes, it is also advisable to wear a face mask, as well as ensuring that the treatment area is well ventilated. Finally the wearing of eye protection is sensible, as many chemicals can cause serious injury if they get into your eyes, and wash your hands after handling chemicals.

Before actually administering any medication, it is vital that the expiry or use-by date should be checked as many chemicals may become more toxic with time or simply be ineffective. When administering any of the treatments described, be sure to allow yourself enough time to do it carefully. Once you are happy that you have made all necessary preparations, you can start treating your koi. But please remember, this should always be under the guidance of, and for some procedures the supervision of, your vet or koi health specialist.

Right: Here an area of infection is receiving topical treatment to encourage healing.

SEDATING YOUR FISH

Times may arise when it is necessary to sedate (anaesthetize) your koi either for treatment or simply to allow easier handling. There are many reasons for sedating your koi including:

1 To calm a koi down and prevent it from struggling and hurting itself.
2 To reduce levels of stress to which a fish is exposed during a procedure.
3 To suppress pain, as many sedating agents have analgesic properties.

It is important to remember that fish probably do experience pain. Anyone who has applied a topical treatment to a open wound is likely to confirm that the koi may flinch or struggle as it is applied, so it must be sensing some form of discomfort, which may be interpreted by us as pain. Therefore it is important that all steps be taken to limit exposure to situations which may

prove painful. This is where sedating agents become a useful tool for both vets and skilled koi health professionals when certain procedures are being undertaken.

The actual process of sedating a fish is one which needs to be learned and thus initially your vet, or koi health specialist, must carry out the procedure. It is only after careful guidance and training from such professionals that you will be able to sedate your own koi successfully. Remember sedating your koi is both a stressful and potentially dangerous procedure for the fish – if too much sedating agent is used and the fish is left in it for too long, it will die. Proper training in the use of sedating agents is vital.

Available Sedating Agents

Numerous sedating agents are available but the most popular for use with koi are MS222 (also known as tricaine methane sulphonate, TMS),

1: Pour pond water into a container to cover the koi.

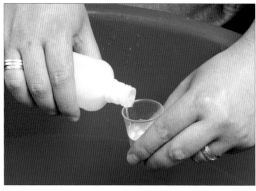
2: Measure out the sedating agent and mix in thoroughly.

3: Place the koi to be sedated into treated water.

4: As the koi is sedated, it will start to list on to its side.

2-phenoxyethanol (or phenoxytol as it is more commonly known), and benzocaine although this must be dissolved in acetone before it can be used. Other sedating agents are available under various brand names depending upon the manufacturer. This type of product must be obtained from a specialist koi dealer or vet.

IMPORTANT NOTICE

If you wish to take a skin scrape or even a swab from a fish, do not sedate it as this may have the effect of reducing the number or activity of any parasites or bacteria present. This may lead you to think that the fish is either clean or harbours only dead parasites or bacteria.

Once suitably trained in using sedating agents by your vet or koi health specialist, the process of sedating (anaesthetizing) a fish is as follows:

1 Measure an amount of pond water into a suitable container, such as a blue bowl large enough for the fish. The amount of water should be kept to a minimum but be enough to completely cover the fish in question.

2 Add enough sedating agent for the volume of water in the bowl, and then mix thoroughly. If using phenoxytol this may take some time because of the oily nature of the liquid. In some cases it may be impossible to dissolve it fully.

3 Once the sedating agent is fully mixed, catch the fish to be treated and move it to the bowl. Once the fish is in the mixture it is vital that you keep a careful watch on its behaviour to ensure that it is removed at the right time. You should see the following signs:

i Fish swims around normally, but reactions start to become slower.

ii Fish stops swimming but is still able to maintain its position in the water.

iii Fish starts to sway from side to side but when upright it can just maintain its position in the water. Gill movement may appear to be slow.

iv Fish starts to lie on its side for longer periods, and appears to be unable to correct its position to upright. However, upon placing your hand in the water to lift the fish, it will try and swim off.

v Fish becomes completely motionless, apart from very slow, and sometimes irregular, gill movement. As this point the fish should not make any attempt to struggle and thus it can be lifted from the bowl ready for treatment or inspection.

If you were to leave the fish in the sedating agent longer, it would eventually lead to the collapse of the brain area that controls swimming and breathing. Thus sedating agents can be used as a humane way of destroying sick or badly damaged fish by simply leaving them in the mixed solution for an hour or two (see page 135).

4 Once you have finished with the sedated fish, it can be returned to the pond or a recovery tank or bowl which should contain the same water as the pond. The water at this point should be heavily oxygenated to aid the recovery process. If using a recovery tank or bowl, the sedated koi should be kept in here until normal breathing is established. When returning the fish to the pond it is advisable to hold the fish in a stream of air to provide a rich supply of oxygen to the gills. The fish can also be gently coaxed back and forth in the water to aid the water flow over the gills. After a couple of minutes the fish should start to show signs of returning to normal, i.e. fin movement, and regular gill movement. At this point you can simply let the fish swim away into the pond. However, keep an eye on it to ensure that it does not get pulled towards any drains or other water outlets while it is regaining its full strength. This process usually takes five to ten minutes, after which it should have completely recovered and returned to normal.

Above left and right: If the sedated koi does not come around by holding it in a stream of oxygen-rich water, use your fingers to open and shut the mouth gently.

Left and above: To speed recovery after sedation, a fish can be gently massaged between the pectoral fins and moved to and fro in the water to improve water flow over the gills.

5 If for some reason, the fish does not come round when put back in the pond the following steps should be taken. While holding the fish in the water by an airstone or venturi, use your fingers to open and close the mouth gently. This helps to push oxygen-rich water over the gills. While doing this move the fish back and forth in the water, as this again helps the flow of water over the gills. You may find it easier to do this by holding the fish belly up, and you may also rub the area between the pectoral fins gently, as this will assist the passage of water over the gills and speed up recovery. Another procedure which might be employed is to use a small pump or pipette to push water into the mouth and over the gills. It may take some time for a fish to come round, but do not give up. Even after one or two hours, in most cases you can get the fish back!

6 Once you have finished with the sedating agent it should be disposed of. Under no circumstances should it be tipped into the pond; instead it should be safely poured to waste.

This information is only given as a guide. It is essential that, until you are properly trained in the use of sedating agents, a vet or koi health specialist must be employed to carry out the procedure. Once you are trained, you will start to notice how different sizes, sexes and varieties of koi react differently when being sedated. It seems logical that a small fish will go over quicker than a larger fish, but if the larger fish has a high fat content, the reverse may be the case as fat absorbs the sedating agent quicker. You will learn how to adjust the dosage to allow for a fish to be sedated quickly, or if dealing with a large number of fish, when to add more sedating agent as each fish absorbs an amount of the original solution from the water. Using sedating agents is an expert skill and professional assistance is required to learn the procedure. However, knowing how to use sedating agents correctly is valuable. It will increase the number of koi that you can coax back to full health, simply because of the extra time that you will have to administer treatments while they are out of the water, and reduce levels of stress and pain that the fish will experience.

Euthanasia

This subject is often not discussed and this leaves the koi keeper having to make the decision on how best to humanely destroy a diseased or injured koi without appropriate guidance. Some unsuitable methods have been practised but these cause the koi much suffering. Unsuitable methods of euthanasia include:

• Slow freezing by placing the fish (in a container of water) in a domestic freezer. Although still routinely practised, it is now considered unacceptable.
• Dropping the fish into very hot/boiling water.
• Leaving the fish to die out of water.
• Snapping the fish's backbone.

Basically, there are two acceptable methods for putting down a fish:

Sedating Agent Overdose

This is the preferred method for most situations. Fish sedating agents such as 2-phenoxyethanol, MS222 and benzocaine can be administered at a lethal overdose level. MS222 is ideal as it dissolves readily in water. Benzocaine on the other hand must be made up as a "stock" solution in alcohol or acetone before further dilution in water to achieve the desired strength. Ideally, ask your local vet or koi health specialist to perform euthanasia on your fish for you.

If, however, you have no option but to perform this procedure yourself, then ensure you administer the correct dosage of anaesthetic, and seek expert guidance if in any doubt. Select a clean water-tight receptacle that is large enough to accommodate the koi comfortably without it having to bend its body or expose any part of its skin above the water. You will need to know the volume of water added to gauge how much anaesthetic to use. You can fill the receptacle with clean pond water. If using an alternative water source, ensure the water is at the same temperature as the koi's pond. The actual dose of sedating agent required will vary according to the size of fish, the water temperature and other factors. As an approximate guide, 0.3 to 0.5g per litre of MS222 will kill most types of fish. Add the koi after mixing in the sedating agent and cover the receptacle. The sedating agent may take 15-20 minutes to kill the fish (sometimes longer) but it is recommended to leave the fish in it for an hour or so, just to be sure. Closely inspect the fish to ensure that all bodily activity (including movement of the gill covers and mouth) has fully ceased before removing it. As a final measure to ensure that the koi is dead, it can be concussed.

Concussion

If no sedating agent is to hand and a fish needs to be put out of its misery quickly, then death by concussion must be considered. The koi is removed from the water and its body (not the head) wrapped in a sheet of soft tissue paper. The fish is gently held on a firm solid surface and its head struck with a heavy object. The aim is to swiftly and fatally damage the brain. This may seem barbaric but is actually fast and effective when performed correctly. If you feel at all uneasy about this procedure, don't attempt it; a half-hearted strike may be unsuccessful and cause even more suffering.

Left: An anaesthetic overdose provides a humane way of killing a severely injured or diseased fish.

TOPICALLY TREATING WOUNDS

This is a treatment that you may need to administer to your koi quite often. Depending on the severity of the condition being treated, you may need to set aside time once or twice a week for it to be done. Topical treatment means treating the area on a fish that is affected rather than treating the whole pond. The advantages are that you are not subjecting the pond to continual treatment which may encourage resistance to certain medications to build up, and also you are treating the affected areas directly which means they can be inspected and cleaned regularly and so should heal quicker. The need for topical treatment may arise after physical damage has occurred, perhaps during spawning or netting, after a parasite infection when scales may have been dislodged by flicking against the pond to try and relieve the irritation, or after a bacterial infection when ulcers have appeared which need to be kept clean while antibiotics take effect.

The actual medication used for the topical treatment varies. Numerous proprietary sprays and ointments are available off the shelf. If these do not appeal, malachite green can be used as an effective topical treatment. It is used undiluted and applied to the area with a cotton-wool bud. When using malachite green gloves should always be worn to protect your hands from exposure to it. An alternative to malachite green, or something to be used in conjunction with it, is propolis spray. This is a natural product made by bees which has beneficial healing and anti-oxidant properties. Propolis can be purchased off the shelf in numerous forms, and is in fact found in many of the products offered commercially for the topical treatment of koi. Propolis can simply be sprayed or painted onto the affected area. If the area to be treated is quite badly damaged or resistant to healing, you may wish to apply malachite green or a proprietary brand of topical treatment first, followed by the propolis. When carrying out topical treatment the following procedure should be followed, in conjunction with any guidelines given on the product being used:

1 If the koi is large or the area to be treated looks like it may require some additional attention before the topical treatment is applied, it is best to anaesthetize the koi. Unless there are adequate reasons, small koi should also be anaesthetized for any invasive procedure.
2 Place the koi on a wet towel, and cover its tail and head with the towel as this will help to keep the fish calm. It may be out of the water for several minutes.
3 Inspect the area and remove or trim any scales

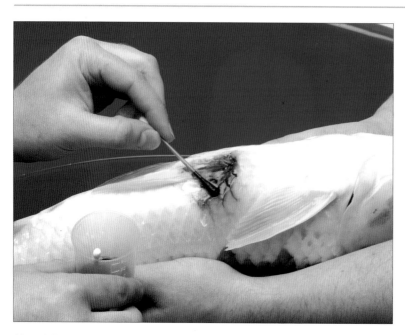

Above left and right: Here a cotton-wool bud is used to apply a topical treatment to an area of infection or physical damage.

Above: Before applying any topical treatment, check that there is no build-up of fluid under the scales. If fluid is present, gently apply the topical treatment without puncturing the skin. This will help to prevent any bacterial infections taking hold in the area.

as required. If any of the scales are raised it is better not to puncture or release the fluid beneath them. The fluid is highly nutritious to many opportunist bacteria, such as *Aeromonas hydrophila*. Even after squeezing out most of the fluid, it will continue to seep, rather like a weeping blister, which encourages secondary infection to take place. Leave the scales intact and seek veterinary advice for an appropriate

antibiotic treatment. If the area to be treated is on the head and there are any loose flaps of skin, cut these away with a sterilized pair of scissors. Stitching by a vet may be an alternative option.
4 Take a cotton-wool bud and gently run it over the affected area to remove any excess fluid or moisture.
5 The site can now be treated with your chosen topical treatment. Care should be taken if working near the mouth, eyes or gills to ensure that the medication does not come into contact with these areas. The gills are perhaps the most vulnerable organs here and extra care should be taken to avoid contact with them.
6 Having finished applying the topical treatment, the koi can be returned to the pond. It will, however, need to be treated again in three to four days' time to keep the area clean and encourage healing. As the affected area starts to heal, the frequency of these treatments can be reduced and eventually stopped.

For quick healing to be achieved, the water in which the koi are kept should be maintained at a temperature of at least 18°C (64°F) or above and in optimum condition. This temperature (or higher) combined with excellent water quality will promote healing and speed up recovery.

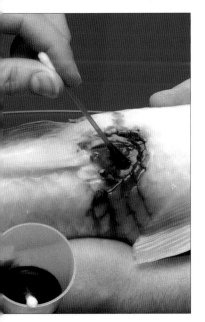

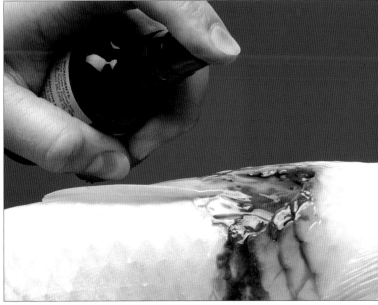

Above: Then propolis spray is applied after the topical treatment.

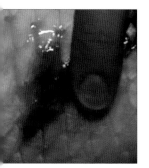

1: To check for the presence of any dead scales, very gently run your finger over the surface of the fish. Dead scales will feel rough to the touch.

2: If scales have to be removed for medical reasons, or are lost due to infection or physical damage, the area will develop scar tissue which will be visible once it has healed.

Removal of Scales

While applying any topical treatments to your koi, it is advisable to inspect the area being treated for other potential problems such as dead or damaged scales. Dead scales have the texture of sandpaper unlike a healthy scale which is smooth to the touch. If the scale has been dead for some time, algae may be found growing on its rough surface. Some scales may just have dead areas. The way you treat them differs and advice should be sought from a vet or koi health specialist as to the best course of action. The easiest way to check for dead or damaged scales is to run your finger very lightly over the surface of the fish. Any scales which are dead will feel rough all over while ones with just dead areas will be smooth except for the places where the scale is starting to die.

Generally the best way to deal with dead scales is to remove them, but this should not be attempted without the guidance of a vet or your koi health specialist, and ideally they should be employed to carry out this procedure until you are properly trained and confident in performing such tasks. If you leave dead scales in place they may become sites of infection. As long as the scale pocket is not damaged when the scale is removed, there is a high chance that the scale should re-grow although this may take many months and even years. If the scale does re-grow,

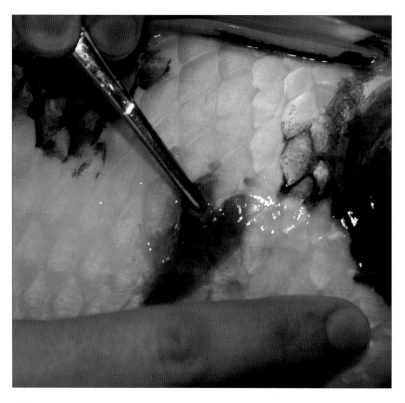

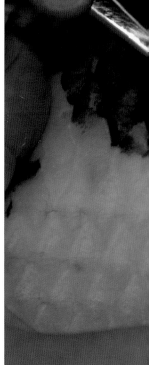

it may not be the same colour as the rest of the area from which it was removed nor exhibit the same intensity of colour.

However, if damage has occurred to the scale pocket through infection or incorrect removal of a dead scale, you may find that new scales do not appear. Instead you are left with a smooth area of skin, which ends up looking like a scar. Some scales may not be completely dead but just contain an area which has died. If this is the case the vet or koi health specialist may choose when treating it to trim the scale by cutting away the dead area rather than removing the whole scale. After either complete removal of a dead scale or trimming of a dead area from a scale, a constant watch should be kept to ensure that no secondary infections occur.

Below (1-3): Here a koi health specialist can be seen removing a dead scale with a sterilized pair of tweezers. This should speed up the recovery of the koi, and may help to prevent further infection from occurring around the site from which the scale was removed.

Treating Fin Damage

When examining areas of fin damage caused by an infection, such as fin rot, through exposure to high ammonia or nitrite levels, or because of physical injury, you may find that the fin or fins are only slightly damaged with small tears and that there are no – or only slight – areas of redness. The best approach here is usually to apply a suitable topical treatment. However, if a fin is very red and inflamed, or the infection has reached the base of the fin and is starting to spread into the body, additional treatment will be required and this must be performed by your vet or local koi health specialist. If the fins are heavily frayed and look inflamed and red, but the problem has not reached the body, it may be necessary for the fins to be trimmed to remove the infection. If the infection has reached the body, there is a chance that some of the bones which are present in the fins may need removing. This is not a procedure for the hobbyist, and a koi health professional or vet should always be employed to carry out any such tasks, and advise on any follow-up treatment.

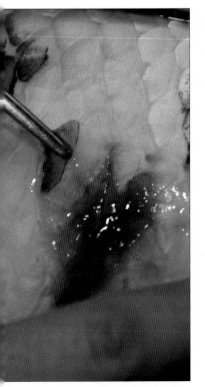

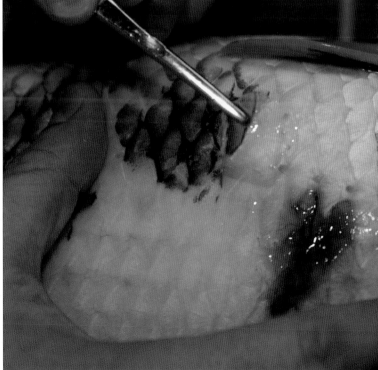

ADMINISTERING ANTIBIOTICS AND INJECTING KOI

Antibiotics are only effective in conjunction with the immune system of the koi. It is preferable to administer these drugs in the feed, although very sick koi may not be feeding and injection is the only alternative. Guidance and advice should always be sought from your local vet or koi health specialist before any decision is made about the use of anti-bacterial medications on sick fish. They may advise that the treatment should be administered in the form of an injection. Very experienced koi keepers may feel confident enough to inject their own fish, but this is potentially a hazardous procedure so we would

antibiotics, and some can have a harmful effect on internal organs, especially the kidneys, and so they should only be used as a last resort after all other avenues of treatment have been exhausted. It is also worth bearing in mind that whenever antibiotics are injected there is also a small associated risk of secondary infection at the injection site as the skin is punctured when the needle is inserted. You must weigh the koi to be injected in order to work out the correct dose, and for this an accurate set of scales is needed. It is easier to weigh your koi if it is sedated as this will both reduce stress

Above: This is a 1ml syringe and is typical of what a vet or koi health specialist will use to inject a sick fish.

recommend that you take advice from your vet or koi health specialist on the best course of action. In many countries antibiotics are not readily available over the counter but must be purchased from a vet who may need to see the fish for legal reasons before he will prescribe the necessary drugs. If you are experienced enough to perform injections yourself, it is a good idea to acquire syringes when picking up the antibiotic as they may also have to be obtained from a vet. Ideally use 0.5ml insulin syringes which come with a pre-attached needle. For bigger koi, which require larger amounts of antibiotics, use a larger syringe with a separate needle.

There is always a risk involved with

and prevent damage from occurring by the fish thrashing about while on the scales. It is vital that either your vet or koi health specialist assists with the sedating procedure unless you have been properly instructed in how to do it yourself. Once the koi is weighed, your vet will be able to calculate the exact dose of the antibiotic required. Although weighing is the only way of getting an accurate measurement, it can be stressful to the koi, and it is time-consuming to do. For this reason some people use the size of the fish to estimate its weight and work out dosages accordingly. If you choose this option, the following size-to-weight ratios may be used. Please note these

Gauging Weight	Below 15cm (6in) or less, fish are too small to inject.		
In order to work out injection dosages, your koi should either be weighed, or you may use the size of the fish to calculate the required dose, using a table like this. These figures are only guides.	Size	Male	Female
	30cm (12in)	0.6kg (1.3lb)	0.8kg (1.8lb)
	35cm (14in)	1kg (2.2lb)	1.25kg (2.75lb)
	45cm (18in)	1.5kg (3.3lb)	1.9kg (4.2lb)
	50cm (20in)	2kg (4.4lb)	2.5 to 3kg (5.5 to 6.6lb)
	55cm (22in)	3kg (6.6lb)	4.5kg (10lb)
	60cm (24in)	4kg (8.8lb)	5.5kg to 6kg (12.1 to 13.2lb)
	70cm (28in)	5kg (11lb)	7kg (15.4lb)

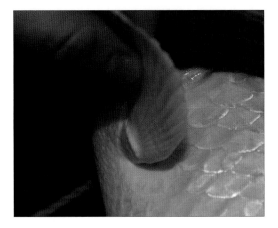

1: Here a koi health specialist can be seen preparing the pectoral muscle ready for injection.

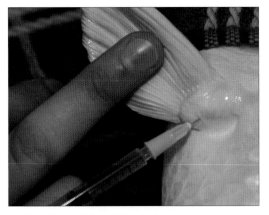

2: The health specialist is now administering the chosen antibiotic via an intra-muscular injection.

3: Finally the injection site is massaged gently to help the dispersion of the medication.

figures are only guides, and if the koi is over- or underweight adjustment may be required.

It is best to perform the injection straight after you have weighed the koi to avoid having to sedate it again. Once the koi is sedated, it can either be laid on a wet towel or held against the side of the bowl in which it was sedated, although this does require some practice to get right and should only be attempted once you are confident at handling koi. An injection can be given in two different ways, although intra-muscular injection is normally the safest. Seek advice from your vet or local koi health specialist as to the best method of application. If inexperienced in this procedure, employ your vet or koi health specialist to do it. The preferred delivery route will be influenced by the type of sedating agent to be given and by the nature and location of the bacterial infection to be treated.

Intra-muscular Injection

This involves the injection of antibiotics into the muscle tissue of your koi, and it can be done in a number of locations. Most veterinarians recommend injection into the dorsal muscle, just below the front ray of the dorsal fin, and the caudal muscle, located between the anal fin and the caudal fin below the lateral line. In order to inject in these positions you have to insert the needle underneath a scale. Any professional would clean the site with an antibacterial before administering the injection. Although an occasional scale may be lost, koi do re-grow scales very rapidly and this is not as damaging as presumed and unlikely to disfigure the koi. These injection sites are in prime locations on a koi and any infection would have a dramatic effect on the aesthetic and monetary value of the fish. If, however, you still choose to inject into one of these locations, be sure that the needle is pointing towards the head of the koi.

Another location for intra-muscular injection is in the pectoral muscle which is located under the pectoral fins. This makes an ideal injection site as it is situated underneath the koi. If any infection of the injection site should occur, there are no scales to be lost, and any scarring will not be easily noticeable. An injection should only be carried out by someone suitably qualified to perform the procedure, such as your vet or local koi health specialist, and the procedure will be conducted in the following way.

1 The koi is sedated.
2 The fish is laid on a wet towel, or held upside down against the side of the bowl in which it was sedated, out of the water.
3 The pectoral fin is rotated so that it is facing away from the body and held there, causing the pectoral muscle to become tense. The injection site is swabbed and cleaned.
4 The loaded syringe is held pointing towards the tail of the koi at a 45-degree angle to the body.
5 The needle is gently pushed into the centre of the taut pectoral muscle; if any firm resistance is felt, the syringe will be pulled out slightly.
6 The contents of the syringe are injected, and then the needle is pulled out.
7 Once the needle is removed, the muscle is gently massaged to encourage dispersion of

the antibiotic. If the pectoral fin is just released and the muscle naturally contracts, some of the antibiotic may be expelled from the injection site.
8 Finally a suitable topical treatment may be applied to the injection site. A slight discharge of blood may occur, but this is no cause for concern.

Intra-peritoneal Injection

This involves injection into the peritoneal space or body cavity located behind the pelvic fins but before the vent and anal fin are reached. With this method of injection there are added risks: if the injection site is wrong, potentially fatal damage to internal organs may occur. Consequently, such injections should only be carried out by, or under the very close guidance of, your vet or local koi health specialist.

Antibiotics and Bactericides

Antibiotics should be stored according to the manufacturer's recommendations, and as a very general rule this is at room temperature below

Below: Here a trained koi health specialist is performing an intra-peritoneal injection. Due to the potential risks involved, this method of injection should only be performed by a vet or trained koi health specialist.

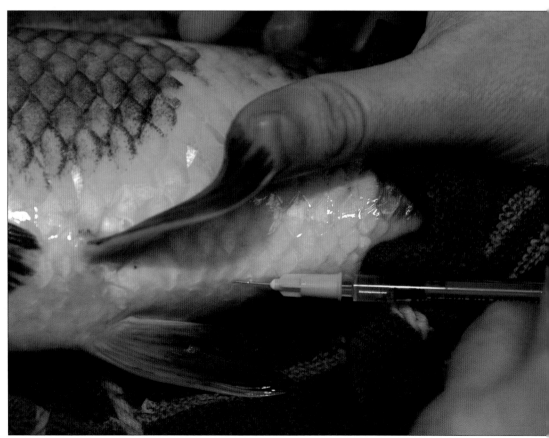

25°C (77°F). A few require refrigeration. It is important to remember that antibiotics have different shelf lives once opened and this shelf life must be strictly observed, along with any other special storage recommendations. Most antibiotics as a rule have to be used within 28 days, but always follow the manufacturer's recommendations. There are numerous anti-bacterials available but really these should only be considered as a last resort after all other possible avenues of treatment have been exhausted, and after guidance from your vet or local koi health specialist.

Before rushing off to your vet for these drugs, it is important to remember that a swab should be taken and sensitivity test carried out and the results shown to your vet so that he can suggest the correct antibiotic or bactericide to use. Guidance should be sought from an experienced fish health consultant regarding the timetable involved in processing the swab when it is sent away for the sensitivity test. In some situations when a fish is extremely ill, the delay in waiting

Above: A fish such as this which is displaying signs of severe emaciation may well be suffering from a bacterial infection that will need to be treated with anti-bacterial medications. In such cases, take advice from a vet.

for the results to come back may prove fatal. In these situations the recommendations of your vet or koi health specialist will be highly valuable. Also it is important to remember that the overuse of antibiotics in itself is an unwise practice, as it may result in antibiotic resistance building up in the bacteria affecting your fish.

To sum up, the use of antibiotics to treat infections should only be considered as a last resort, and then only after guidance from a vet. It pays to take time to discover a vet near you who has experience of fish health problems. Along with advice on antibiotics and medications, he or she will be able to offer many other important professional services, such as x-rays and ultrasound to look for organ abnormalities and tumours, endoscopy, histological examinations, post-mortems and various surgical procedures.

TREATMENT BATHS AND DIPS

Some of the diseases discussed in section 3 can readily be treated with baths or dips, usually either a salt bath or a potassium permanganate dip. When using a bath or dip treatment, do take care to follow the guidelines described below; if unsure, seek advice.

Salt Baths

These are an ideal way to treat numerous infections and may be administered to a single fish or to all the koi in the pond. It is vital to use the correct salt – this should be cooking salt not table salt. Table salt contains bleaching agents and other impurities which may harm your koi. Not only is salt suitable for bath treatments, but it can also be added to the pond in very low temperatures when other medications prove ineffective, and when dealing with a condition that responds well to such pond treatment.

As a general rule, however, avoid adding salt to your pond as it will limit your use of other medications, such as formalin. Salt does not degrade in the pond and the only way it can be removed is through water changes. So salt is best suited to bath treatments for individual koi. The strength of the salt bath will vary depending upon the condition being treated, although as a guide 22g per litre of water (3.5oz per gallon) for 10 minutes is a standard dose. This is how to give a salt bath.

1 Measure enough water into a viewing bowl or other suitable container to cover the koi to be treated. This should be pond water not tapwater! Ideally put in an airstone as well to provide the fish with sufficient oxygen in the bath.
2 Measure out the right amount of salt in relation to the volume of water in the bowl.

1: Add the required amount of salt to some pond water.

2: Having added the salt, mix it thoroughly.

3: Start timing as soon as the fish enters the water.

4: Fish may list in the water while being treated.

3 Thoroughly mix and dissolve the salt.

4 Transfer the koi into the salt bath. As soon as it is in the water, be sure to have a stopwatch to hand to time how long it should remain in the bath. Ten minutes is the maximum time that fish should spend in the bath.

5 While koi are in a salt bath they may appear to be stressed and many may float on their sides as if they were dead. Do not be alarmed by this, and leave the fish for the duration of the bath. This is purely the effect of the salt and once the koi is taken out of the bath it should make a quick recovery. If the koi becomes distressed, such as becoming disorientated, with very shallow breathing, immediately remove it from the salt bath and return it to the pond.

6 After the designated time remove the koi from the salt bath and return it to the pond. Pour the water used for the salt bath to waste.

Potassium Permanganate Dips

These are much harsher than salt baths and are generally used when a specific problem, such as a parasite, is identified. Although beneficial, in alkaline water with a high pH, potassium permanganate becomes toxic. Manganese dioxide can form, which coats the gills and kills the koi. Take adequate steps to protect yourself and the surrounding area from being stained by the chemical. As with salt baths, the strength of the dip required depends upon what you are dipping for, but a standard dip for most parasite infections is 100mg of potassium permanganate per litre (0.016oz per gallon) for five minutes. This is how to give the dip:

1 Measure enough water into a viewing bowl or other suitable container to cover the koi to be treated. This should be pond water not tapwater! As potassium permanganate oxidizes in water, it is recommended that airstones be added before the chemical is mixed into the water.

2 Measure out the right amount of potassium permanganate for the dip. To help reduce the oxidizing properties of the chemical, you may wish to mix it first in warm or hot water and add an airstone to this mixture. Leave it for 20 minutes before adding it to the bath.

3 Transfer the koi to be treated into the dip; if airstones are to hand add these to the dip while the koi are in it. Again, use a stopwatch to time how long the fish is in the water. This chemical

1: A measured amount of potassium permanganate is being poured into a treatment bath.

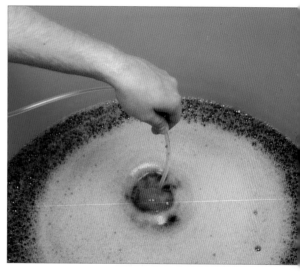

2: An airstone should be added to the dip at least 20 minutes before the koi is placed into it.

stains so watch out for any splashing.

4 While koi are in the dip they may appear to be very stressed and many may float on their sides as if they are dead. Don't be alarmed, just leave them for the duration of the dip. This is purely the effect of the chemical and once the koi are taken out of the dip they should make a quick recovery. However, weak koi may not survive the dip, but in such cases the chances are that these koi would probably have died shortly, even if the dip had not been used.

5 After the designated time has passed, remove the koi from the dip and return it to the pond. Wear protective clothing and pour the water used for the salt bath to waste.

POND, TOPICAL AND ALTERNATIVE TREATMENTS

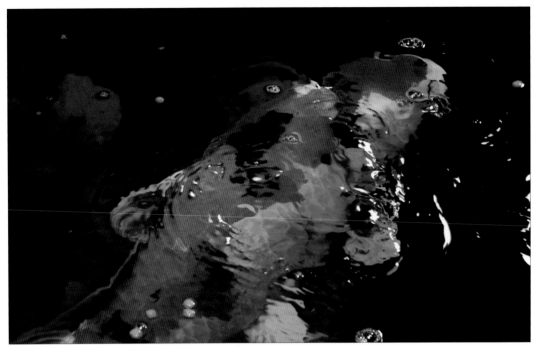

Many chemicals mentioned in section 3 can be used to treat the whole pond, and/or used as a topical treatment. These commonly come as either a liquid or in powder form. Some of the treatments covered here are not explicitly mentioned in section 3, but they provide a useful alternative to those which are recommended, and may be preferred by some koi keepers who are already familiar with them. Of course, many more treatments than are mentioned here are commercially available. In all cases it pays to take advice from your koi health specialist or vet regarding the most effective treatment to use.

Whether you are treating the whole pond or simply topically treating a wound, it is vital that precautions are taken regarding the safe handling of the product. This may involve wearing gloves or working in a well-ventilated area, or even wearing a face mask. Once these precautionary steps are taken to protect yourself, it is critical that the correct amount of medication is measured out for the required treatment. If the whole pond is being dosed, you must know its volume in litres or gallons. Ideally this figure should have been obtained when the pond was built and filled, normally by using a flow meter. However, if you have not done this or perhaps have moved into a house with an existing pond, the volume of water will have to be worked out as best you can. The easiest way to do this if the pond is of a square or rectangular shape is to measure the length and width, take an average depth, multiply these figures together to give a measurement in cubic feet and multiply this by 6.23 to give a figure in

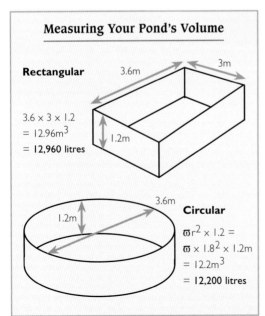

Measuring Your Pond's Volume

Rectangular   3.6m   3m

$3.6 \times 3 \times 1.2$
$= 12.96m^3$ 1.2m
$= 12,960$ litres

3.6m

Circular

$\varpi r^2 \times 1.2 =$
$\varpi \times 1.8^2 \times 1.2m$
$= 12.2m^3$
$= 12,200$ litres

1.2m

1: Here a clay additive for the pond is being mixed. (As clay is harmless, gloves are not required.)

2: Mix until the clay is fully dissolved. If mixing chemicals, **never** use your bare hands – always wear gloves.

3: Once the clay is fully dissolved, it can be evenly poured over the surface of the pond.

Imperial gallons. Multiply this by 4.546 to express the figure in litres. If you are working with metric measurements, the figure in cubic metres should be multiplied by 1000 to give a volume in litres. If the pond is an irregular shape, it is best to divide it up into zones, and work out the volume for each zone using the same method, then add all of these together. This can also be done if the pond has numerous areas at different depths. If the pond is round, you will need to measure the radius and use the formula of ϖ (3.1416) x radius squared, multiply this by the average depth and multiply this figure by 6.23 to give a gallons measurement.

Dispensing Medications

With this information you are now ready to start measuring out the required medications. If working with liquid an accurate measuring cylinder is vital, and it is a good idea to own several in different sizes, as the amount of medication will differ from chemical to chemical. A larger measuring jug is also a good investment for those chemicals which have a higher dose rate or if you have a very large pond. Not all pond treatments are liquid, so a good accurate set of gram scales is an essential piece of equipment. These should be able to measure out down to 0.1 of a gram as in many cases you are dealing with very small amounts of powder, and thus 0.1 of a gram can make a big difference. Scales which are this accurate can be hard to find and are expensive but your local koi dealer should be able to help you to source a set. If this is not the case, ask your koi health specialist or local koi dealer to measure treatments for you as generally they own a pair of very accurate scales for their own use.

Before adding medication to the pond turn off any UV filters to stop them from breaking down the medication. These can be turned on again after 24 hours. If you run a sand or bead filter this should be put on to recirculate as well to prevent it from filtering the treatment out before it has a chance to do its job. These can be reset to normal filter mode after 24 hours as the treatment will usually have done its job by then, unless the directions for the medication state differently.

When adding a treatment to the pond it is best (unless otherwise stated or advised by your local koi specialist) to mix it with some pond water, then evenly distribute it over the pond's surface. At the same time try to increase the level of aeration in the pond, as many treatments reduce oxygen levels. This will help to reduce stress. While the treatment is active within the pond, it is also advisable to avoid feeding for 24 to 48 hours unless the treatment advises differently. If you are using the treatment for a topical application, it is normally only necessary to measure out enough medication to apply it to the area in question, and this can be simply done in a suitable container.

It is always advisable to store the treatments described here in a cool dark cupboard, unless the manufacturer advises differently. Ensure that they are accurately labelled and stored well away from children and animals. It is good practice to check the use-by dates on a regular basis, and also to check each chemical for any visible signs that it may have gone off. Formalin, for instance, can go off, and the easiest way to check for this is to ensure that there are no white crystals or lumps present in what should be a clear liquid. If any are seen, dispose of the substance immediately.

Acriflavine
Crystals, but it is more commonly sold now as a fluorescent green/yellow liquid. Used mainly to treat bacterial, fungus, and mild parasitic infections.
Dose rate: This will vary depending upon the solution mix, as different manufacturers use different strength solutions.
Topical use: Not appropriate.
Bath use: Not appropriate.
Frequency: As required, and as recommended by the manufacturer.
Notes: Acriflavine is less widely used nowadays due to the limited levels of success that it offers. If a stronger alternative is required, proflavin hemisulphate can be used. However, this is expensive, and it will be very damaging to the beneficial bacteria in your biological filters.

Chloramine T
A white powder. A very effective treatment for both parasitic and bacterial conditions, especially bacterial infections of the gills. Advice should always be sought before using it as dose rates differ depending upon the pH of your pond.
Dose rate: Varies depending upon the condition being treated. Minimum effective dose is 1g per 1000 litres (220gallons), although a dose five times stronger than this (or even stronger) can be used in extreme conditions. Seek advice before using higher dose rates, as it can be toxic to koi.

Above: It is vital that treatments are measured accurately to avoid any risk of a potential overdose.

Topical use: Not appropriate.
Bath use: Not appropriate.
Frequency: As directed, varies depending upon the condition being treated.
Notes: Before using Chloramine T you must check your pH and water hardness as it is very toxic in soft water. If you live in an area with a pH of below 7 and have soft water, Chloramine T is not the best medication to use and advice should be sought from your vet or koi health specialist on the best course of action.

Formalin
Clear liquid. Normally used with malachite green as an all-round anti-parasite and general pond treatment.
Dose rate: Varies depending upon solution strength and manufacturer.
Topical use: Not appropriate.
Bath use: Can be used as a bath; the dose rate varies depending upon solution strength and manufacturer.
Frequency: When using as a pond treatment with malachite green, once every five to seven days up to a maximum of three times.
Notes: Formalin goes off, and the easiest way to check for this is to ensure that there are no

white crystals or lumps present in what should be a clear liquid. If any are seen, dispose of the substance immediately. When handling formalin care should be taken to avoid skin contact and breathing in any fumes. Formalin should never under any circumstances be used with potassium permanganate. If using in a pond which already contains salt, advice should always be sought from your koi health specialist before applying this treatment.

Hydrogen peroxide 3 per cent solution

Clear liquid. Used as a cauterizing agent (will seal the wound and stop bleeding) as a topical treatment.
Dose rate: Not appropriate.
Topical use: Apply to the affected area with a cotton-wool bud. Should only be used when advised to do so by a vet or health specialist due to the hazardous and potentially harmful nature of this product.
Bath use: Not appropriate.
Frequency: As advised.
Notes: A very powerful treatment which should only be used when directed, as it can cause severe tissue damage which may hinder rather then help the healing process if used inappropriately. When treating, avoid getting this substance on the gills or other sensitive areas, such as the eyes.

Malachite green

Crystals but commonly sold now as a very dark green liquid in its dissolved form. Should never be handled neat in its crystal or powder form as it is highly toxic if breathed in. An excellent treatment for whitespot if used alone, plus a topical treatment. When used with formalin an excellent all-round general pond treatment.
Dose rate: Varies depending upon solution strength and manufacturer.
Topical use: Simply apply to the area with a cotton-wool bud as required.
Bath use: Not appropriate.
Frequency: When using as a pond treatment once every five to seven days up to a maximum of three times. Same applies if using with formalin.
Notes: Always wear gloves when working with liquid malachite green.

Potassium permanganate

Can be brought as liquid or in crystal form

Above: Common treatments showing their form and colour.

– purple in colour. Mainly used as a pond or bath treatment for numerous diseases, particularly parasites or bacterial infection. Can also be used as a topical treatment where it acts as a cauterizer; but it is potentially harmful and should only be used when advised to do so by your koi health professional or vet.
Dose rate: Numerous dose rates are used. The standard dose is 1.5g per 1000 litres (220 gallons), but as potassium permanganate is rendered inactivate by organic debris in the pond, dose rates may vary depending upon the levels of organic debris present. The more debris present, the less effective the treatment will be.
Topical use: Mix a very small amount of potassium permanganate crystals with water to create a slurry, which can be applied with a cotton-wool bud. Acts a cauterizer so should only be used when bleeding occurs as it will seal the skin and stop the bleeding. Should only be used when advised by a vet or koi health specialist.
Bath use: 100mg per litre (0.016oz per gallon) for five minutes. This is very strong but effective. If treating weak fish you should expect losses, but those which survive the bath will probably make a full recovery. Don't exceed five minutes!
Frequency: As a pond treatment it can be repeated up to a maximum of three times leaving five to seven days between each dose. As a bath, use as required, but avoid excessive use.
Notes: Potassium permanganate should never be mixed with formalin. If using in a pond which already contains salt, advice should always be sought from your koi health specialist before applying this treatment.

Propolis

A topical treatment and food additive. Generally available in spray form or as a liquid with a brush for easy application. If using as a food additive, different concentrations are available.

Dose rate: Not appropriate.

Topical use: Apply to the affected area as required.

Bath use: Not appropriate.

Frequency: As required.

Notes: A natural product produced by bees. Can be mixed daily with food to improve your koi's immune system and overall health.

Above: Propolis spray makes an excellent topical treatment either on its own for small superficial damage or used in conjunction with a proprietary topical treatment for more severe areas of infection.

Above: Two different versions of propolis – the smaller bottle is a spray for wound treatment while the larger bottle is a food additive.

Salt

White granules. Used to reduce osmotic pressure as a pond treatment and help reduce stress levels. Also used as a bath to reduce parasite levels.

Dose rate: The amount of salt used varies from 7g (0.25oz) per 4.5 litres (1gallon) up to 6kg (13.2lb) per 1000 litres (220 gallons).

Topical use: Not appropriate.

Bath use: Strength of bath varies, standard doses are between 85-100g (3-3.5oz) per 4.5 litres (1 gallon) for ten minutes.

Frequency: Pond treatment will normally be a one-off, unless you want to maintain a specific level, in which case salt will need to be added after water changes. To determine how much is required, the salinity will need to be monitored using a hydrometer or salt tester, but these only offer a limited level of accuracy and ideally an electronic salt tester or TDS (total dissolved solids) meter should used. If given as a bath, everything depends upon the condition being treated. Generally one bath a day for three days is the maximum.

Notes: Salt is an effective treatment at all water temperatures. The disadvantage of adding salt to your pond is that it will not biodegrade and the concentration can only be reduced through water changes, while continued salt use can lead to salt-tolerant parasites as is thought to be the case with trichodinids. If you wish to use other medications when salt is also present in the pond, advice should always be sought from your vet or koi health specialist.

General Advice

The use of medication in your pond can have an effect on the filtration system so it is vital after any course of treatment to perform extra water tests, and to take the necessary steps to improve the water quality. You must follow the guidelines

Above: Salt is an excellent bath treatment, and is easily measured.

drastic behavioural changes in your koi, such as gasping, jumping, erratic swimming movements, or at the other extreme they may start to float or sink to the bottom and list from side to side. If this does happen, your first step should be to start to change water and increase aeration and continue to do so until your koi start to behave normally again. For routine pond maintenance no more than 10 per cent of the water should be changed at any one time to avoid causing stress to the fish and, if the pond is heated, lowering the temperature too rapidly. However, in the event of an overdose, it may be necessary to change larger volumes of water to dilute the quantity of medication present more quickly. Even then, try to avoid changing more than 40 to 50 per cent of the water at a time.

printed on the packaging, or seek professional advice from your vet or local koi specialist before dosing your pond. In the very unfortunate event of overdosing a pond, you will soon notice

The treatments described above are only a small selection of what are available for the koi keeper. Most manufacturers of koi products produce their own range of medications which can be used effectively to treat most conditions which the koi keeper will experience. This is helpful as not all of the treatments mentioned are readily available worldwide, as different countries have different regulations regarding acceptable medications.

This has led to increased interest in alternative approaches, and the use of probiotics. These are not medications as such, but friendly bacteria which are introduced into the pond via a liquid product. In very simple terms these friendly bacteria eat and live on the harmful bacteria or their food source and so help to create a healthier environment for the koi. Probiotic foods have also become popular. It must be stressed that many of these new alternative approaches are still in their infancy and thus have yet to be fully evaluated for their claimed benefit. Food additives such as propolis and specific vitamin supplements are also popular, although if a good quality branded food is purchased in the first place, their use is not essential. All of these approaches aim to improve your koi's overall health and so reduce the incidence of disease, and hence the need to use medication.

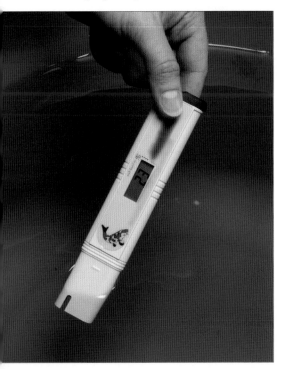

Left: If salt is added to the pond, it is vital that the strength of salinity is known as some treatments cannot be used with salt. Here a digital salt meter is being used to check the salt level in the water.

SHOULD YOU QUARANTINE OR NOT?

Quarantining new koi purchases is a matter of personal choice. It is also a subject that arouses much discussion. You can argue that if you buy from a reputable dealer who has already kept the koi in quarantine for some time before sale, an additional quarantine period is not required. However, then you must make sure that koi are only purchased from a single, or very limited number of, sources – otherwise the risk of problems arising increases dramatically. Others argue that you should quarantine all new purchases for between four and eight weeks as the stress of transportation and the change in water quality from the dealer's ponds to your pond alone creates a need for an additional quarantine period. This argument is valid, but to be effective the quarantine facilities need to replicate your main pond; if that is not the case, when the koi are finally moved into your main pond they will undergo the same stresses that they would have when purchased.

it infects your pond, quarantining is a decision to which more people are saying 'Yes'. The problem with quarantining is simply this: in most instances the factors which contribute to a koi succumbing to serious diseases are stress, poor water quality, and environmental change. Now while in an ideal world every koi would be quarantined in optimum conditions, the majority of quarantine tanks are actually small units with limited filtration, often located inside a shed or garage. This causes two problems. Firstly, the size and set-up of a typical quarantine system does not create the ideal environment for your koi, and it is very difficult to keep the filters mature, as you do not want to keep fish in the quarantine tank all the time as these may develop their own infections which could be passed on to newly purchased stock. Secondly, because of the location and positioning of most quarantine systems, they tend to take second place to your main pond, and they suffer as a consequence.

How can this problem be overcome? The ideal solution is to have a large quarantine system – a second pond in fact, not necessarily landscaped and decorated, but holding at least 4500 litres (1000 gallons) of water,

Above: This koi is suffering from an attack of fungus. The virtue of running a quarantine system is that you can isolate such fish to lessen the chances of disease transmission.

heated and containing all the filtration equipment that you would consider putting in your main pond, and perhaps even some extra units, such as ozone and UV sterilizers. Unfortunately this is beyond most people's budgets and they just do not have room for such a set-up. Thus a compromise should be sought and this can take the form of a quarantine facility, but one which is

Should you quarantine or not? There is no definitive answer to this, but as more people become aware of devastating diseases such as KHV that can destroy all your existing stock if

not necessarily for isolating newly purchased koi. Instead this facility may be used to move koi from your main pond when they succumb to infection and require specialist treatment, such as very warm water, regular injections, or the introduction of medication which may be detrimental to the filter system. If, however, a fish only has physical damage or an isolated infection which requires topical treatment, it is better to leave your koi in the main pond as they will recover more quickly there, and only move them out as a last resort.

When buying new koi, you should be able to trust your dealer if the business is reputable. The dealer should be able to tell you when the fish were imported, what they have been treated with, any problems which they might have experienced, if they have been temperature-tested for KHV, and if the farm they have come from has been tested for KHV. Ultimately a reputable koi dealer will not knowingly sell diseased or infected stock, as the consequences for the business would be catastrophic. So when choosing your principal source for new koi, take time to get to know the staff and look at the stock. You should not see large numbers of

Above: This well-maintained quarantine system belongs to a private koi keeper. It has all the filtration equipment that you would normally find associated with a main koi pond.

dead koi in the display ponds, nor lots of fish with evident signs of injury or damage. The koi which are for sale should all be lively and readily looking for food, and appear in overall good health. Try to speak to regular customers and see how they have found koi purchased from this dealer. Ask if they have ever quarantined any purchases and, if not, have they ever experienced health problems.

With all this information to hand you should be able to buy your koi in confidence and be happy to introduce them to your main pond without the need for any extended quarantine programme. Your own quarantine facilities can be used more as a treatment-cum-isolation tank for any serious infections which need separate treatment. In the end there is no "right" answer. It is a matter of personal preference guided by the advice of your local koi dealer who will help you to make your final decision.

CONCLUSION

Koi health should not be a daunting subject. Many of the diseases described in this book can easily be prevented by careful, thorough system maintenance and good husbandry. The single biggest cause of disease is stress and if this can be kept to a minimum, the outbreak of health problems will be dramatically reduced. The largest stress factors in your koi pond are poor water quality, temperature fluctuation, overstocking, bad handling, frequent netting, and poorly maintained filtration systems. If these factors can be closely monitored and maintained in optimum conditions, you are well on your way to creating a healthy and stable environment in which your koi will thrive.

Diseases do unfortunately occur, no matter how well your pond is maintained. However, if correct identification is made quickly and the appropriate course of treatment followed, the common complaints which the koi keeper will experience can be cleared up quickly and easily. Of course, there are some conditions which do require more specialist treatment and are harder to cure; however, these will only affect a handful of koi keepers, and so should not be a cause for general alarm. It is all too easy to convince yourself that your koi have a rare and serious condition, but the chances are that they do not.

When using this book to identify symptoms, always start with the common problems, such as parasites or mild bacterial infections. A lot of the diseases which affect koi exhibit very similar symptoms, but it makes sense to start with the obvious diseases in trying to make an exact diagnosis. Correct identification is the key to the treatment and successful cure of all koi diseases, and so as much time and effort should be put into making the diagnosis as into the treatment itself. The trap here is that you will do the easy things, like taking a skin scrape which requires little time

or expense, but forget to take a swab, which has to be analysed by an independent laboratory for a charge. It is only when all the necessary tests have been carried out that an accurate diagnosis can be made, and the correct course of treatment selected.

As you come to the end of this book, you should now be able not only to identify the causes of many of the health

The goal of all koi keepers – beautiful and healthy fish swimming gracefully for our pleasure.

problems covered, but also be able to treat them accurately and thoroughly and keep losses to a minimum. Despite this, it is always advisable to seek the opinion of a suitably qualified vet or koi specialist before administering any treatments. Hopefully, you have been persuaded to buy a microscope and install heating in your pond. These items are costly, which, unfortunately,

tends to deter some people, but the same people will quite happily spend a similar amount of money on a new fish. Remember, the investment made in heating and a good microscope will easily pay for itself in time, as these items will help you to keep the pond healthier, identify any problems quicker, and ultimately save more koi when disease does occur.

Finally, perhaps the most important point to stress is that everyone has a different opinion and different experiences. What works for one person will not always work for another. When dealing with the health of your koi, it is vital that you decide whose advice you are going to follow, and stick to it. Far too often fishkeepers will take advice from numerous sources and then mix up the recommendations, following different advice from different people. This is asking for trouble. If you decide to use the treatment methods outlined in this book, it is vital that they are followed as described. Don't use them in conjunction with other treatments as this could cause complications. Conversely, if you are currently pursuing a course of treatment and want to change to one of the treatments suggested in this book, seek advice before doing so. When you have fully assimilated the advice offered in this book, you should be in a much better position than you were beforehand to provide your koi with a high standard of health care, but always discuss things in full with your local koi specialist, and do not be afraid to ask for their help and advice. Finally keep the book by you as a source of reference and enjoy your koi!

INDEX

Further Reading

Andrews, Dr Chris, Exell, Adrian and Carrington, Dr Neville *The Interpet Manual of Fish Health* (Interpet Publishing, 2002)

Hickling, Steve, Martin, Mick, Brewster, Bernice and Fletcher, Nick *Koi – A colourful and comprehensive celebration of these beautiful ornamental fish* (Interpet Publishing, 2002)

Jepson, Lance *Koi Medicine* (TFH Kingdom Books, 2001)

Johnson, Erik L. *Koi Health and Disease* (1997)

Kuroki, Takeo *Manual to Nishikigoi* (Shuji Fujita, 1981)

Lammens, Maarten *The Koi Doctor – your guide to keeping healthy koi* (A-Publishing/KINDAI bvba, 2004)

Nishikigoi Manual – Nishikigoi varieties and their diseases (published in Japanese with an English parallel text) (All-Japan Nishikigoi Promotion Association in association with Kinsai Shuppan)

Post, Dr George *Revised and Expanded Textbook of Fish Health* (TFH Publications, 1987)

Rinko magazine (published monthly by Shin Nippon Kyoiku Tosho Co., Ltd, Japan)

Untergasser, Dieter *Handbook of Fish Diseases* (TFH Publications, 1989)

Picture Credits